I0493840

<u>Disclaimer</u>

Book Title: Measurements and Models for the Wireless Channel in a Ground-Based Urban Setting in Two Public Safety Frequency Bands

Book Author: William F. Young; Catherine A. Remley; David W. Matolak; Qian Zhang; Christopher L. Holloway; Chriss A. Grosvenor; Camillo A. Gentile; Galen H. Koepke; Qiong Wu;

Book Abstract: We report on measured peer-to-peer (ground-based) wireless channel characteristics for an urban environment in two public safety frequency bands. Results are based upon measurements taken in Denver in June 2009. The public safety bands we investigated are the 700 MHz and 4.9 GHz bands, both intended for public safety and emergency-response applications. Our study of the urban environment in these bands included an estimation of the distributions of both the number of multipath components and their delays. Our measurements employed a vector network analyzer, from which both path loss and delay dispersion characteristics were obtained for link distances up to approximately 100 m. Log-distance models for path loss are presented, and dispersive channel models are also described. Our dispersive channel models employ a statistical algorithm for the number of multipath components, previously used only in indoor settings. By employing a transmit-antenna positioner, we introduced spatial diversity into the measurement system, which enabled analysis of the dispersion characteristics of the angle of departure, also new for this ground-to-ground channel. The channel models should be useful for public safety communication system design and development.

Citation: NIST TN - 1557

Keywords: attenuation; delay spread; emergency responders; multipath; public safety; radio communications; radio propagation experiments; transfer function; urban canyon; wireless communications.

National Institute of Standards and Technology
Technology Administration, U.S. Department of Commerce

NIST Technical Note 1557

Measurements and Models for the Wireless Channel in a Ground-Based Urban Setting in Two Public Safety Frequency Bands

William F. Young
Kate A. Remley
David W. Matolak
Qian Zhang
Christopher L. Holloway
Chriss Grosvenor
Camillo Gentile
Galen Koepke
Qiong Wu

NIST Technical Note 1557

Measurements and Models for the Wireless Channel in a Ground-Based Urban Setting in Two Public Safety Frequency Bands

William F. Young
Kate A. Remley
David W. Matolak
Qian Zhang
Christopher L. Holloway
Chriss Grosvenor
Camillo Gentile
Galen Koepke
Qiong Wu

Electromagnetics Division
National Institute of Standards and Technology
325 Broadway
Boulder, CO 80305

January 2011

U.S. Department of Commerce
Gary Locke, Secretary

National Institute of Standards and Technology
Patrick D. Gallagher, Director

Certain commercial entities, equipment, or materials may be
identified in this document in order to describe an experimental
procedure or concept adequately. Such identification is not
intended to imply recommendation or endorsement by the
National Institute of Standards and Technology, nor is it
intended to imply that the entities, materials, or equipment
are necessarily the best available for the purpose.

**National Institute of Standards and Technology Technical Note 1557
Natl. Inst. Stand. Technol. Tech. Note 1557, 67 pages (January 2011)
CODEN: NTNOEF**

U.S. Government Printing Office
Washington: 2005

For sale by the Superintendent of Documents, U.S. Government Printing Office
Internet bookstore: gpo.gov   Phone: 202-512-1800   Fax: 202-512-2250
Mail: Stop SSOP, Washington, DC 20402-0001

Contents

Executive Summary

This is another in a series of NIST technical notes (TN) on propagation of radio signals for emergency-responder communications. Previous technical notes investigated propagation into large building structures (apartment complex, hotel, office buildings, sports stadium, shopping mall, etc). Three NIST Tech Notes (NIST TN 1540, NIST TN 1541, and NIST TN 1542) described experiments related to radio propagation in a structure before, during, and after implosion. Two subsequent Tech Notes (NIST TN 1545 and NIST TN 1546) focused exclusively on RF propagation into large buildings, with no implosion results. Those reports were intended to give emergency responders and system designers a better understanding of what to expect from the radio-propagation environment in disaster situations. The overall goal of this project is to create a large, public-domain data set describing the attenuation and variability of radio signals in various building types and environments in the public safety frequency bands. These studies have been funded by the NIST Public Safety Research Laboratory within the Office of Law Enforcement Standards.

Because the Federal Communications Commission (FCC) has recently opened spectrum between 764 MHz and 776 MHz for public safety applications, in this report additional measurements were carried out in the 750 MHz frequency band in a dense urban environment (sometimes called an "urban canyon"). We also conducted measurements in this urban environment in the 4900 MHz public safety band. (4940 MHz to 4990 MHz represents another licensed public safety band.) These measurements were all "ground-to-ground" with transmitters and receivers at ground level and pedestrian-height antennas to mimic communications between an incident command post and a responder on foot.

This report describes measurements conducted in the urban environment with a vector network analyzer (VNA). The VNA yielded samples of the channel frequency response for multiple transmitter and receiver locations, and from these frequency responses (or "channel transfer functions"), channel impulse responses were derived. The channel impulse responses were analyzed and delay dispersion statistics were gathered for two types of channels: (1) those with a line of sight (LOS) between transmitter and receiver; (2) those that were non-line of sight (NLOS). For a given channel type, the average delay spreads in the two bands—the 700 MHz and 4900 MHz bands—were found to be less than 120 ns for LOS for distances up to 80 m, and between 50 and 240 ns for NLOS conditions for distances between 50 and 140 m.

We also gathered statistics on the distribution of multipath components: their number, their delays, and their amplitudes. These distributions can be used as parameters for channel models, which can be used to assess the performance of any type of communication system used in these frequency bands in these environments.

Measurements and Models for the Wireless Channel in a Ground-Based Urban Setting in Two Public Safety Frequency Bands

William F. Young, Kate A. Remley, David W. Matolak, Qian Zhang, Christopher L. Holloway, Chriss Grosvenor, Camillo Gentile, Galen Koepke, Qiong Wu

Electromagnetics Division
National Institute of Standards and Technology
325 Broadway, Boulder, CO 80305

We report on measured peer-to-peer (ground-based) wireless channel characteristics for an urban environment in two public safety frequency bands. Results are based upon measurements taken in Denver in June 2009. The public safety bands we investigated are the 700 MHz and 4.9 GHz bands, both intended for public safety and emergency-response applications. Our study of the urban environment in these bands included an estimation of the distributions of both the number of multipath components and their delays. Our measurements employed a vector network analyzer, from which both path loss and delay dispersion characteristics were obtained for link distances up to approximately 100 m. Log-distance models for path loss are presented, and dispersive channel models are also described. Our dispersive channel models employ a statistical algorithm for the number of multipath components, previously used only in indoor settings. By employing a transmit-antenna positioner, we introduced spatial diversity into the measurement system, which enabled analysis of the dispersion characteristics of the angle of departure, also new for this ground-to-ground channel. The channel models should be useful for public safety communication system design and development.

Key words: attenuation; delay spread; emergency responders; multipath; public safety; radio communications; radio propagation experiments; transfer function; urban canyon; wireless communications.

1. Introduction

When emergency responders work in an urban environment, radio communication to individuals on the street or around corners is often impaired. Mobile-radio and cell-phone signal strength is reduced due to attenuation caused by large buildings, and these buildings also cause signals to reflect and travel multiple paths from transmitter to receiver [1-8]. This multipath propagation can cause severe signal distortion, which can significantly degrade performance of a communication system operating in this environment. In addition, when emergency responders set up a communication system on scene, antenna heights will often be low, on the order of pedestrian heights, and this can degrade coverage.

Wireless communications for public safety authorities are seeing increased attention [9], [10]. Several bands in the 700 MHz spectrum, formerly allocated to television broadcast, have been re-allocated for public safety applications nationwide, and a band in the 4900 MHz spectrum also has been recently allocated. In the 700 MHz band, two 12 MHz blocks are available from 764 MHz to 776 MHz and 794 MHz to 806 MHz, whereas in the 4900 MHz band, 50 MHz is available from 4940 MHz to 4990 MHz.

Public safety communications have traditionally been "narrowband," with voice the primary service. Channel allocations of 6.25 kHz, 12 kHz, and 25 kHz have been used for many years. The use of new, wider-band services has been gaining popularity for applications such as video, geolocation, etc., and this has initiated development of wider-band air interface standards, such as the so-called P34 standard, originally developed by the Association of Public Safety Communications Officials (APCO), now part of the Telecommunications Industries Association (TIA)[11]. With the tremendous growth of wireless local- and metropolitan-area networks which use the IEEE 802.11 and 802.16 (WiMAX, [12]) standard technologies, as well as cellular technologies such as the 3GPP's Long Term Evolution (LTE) standard, the public safety community is likely to employ one or more of these technologies for reasons of reliability and economy. Typical signal bandwidths for these technologies are 1.25 MHz, 5 MHz, 10 MHz, and 20 MHz.

For any wireless communication system to operate reliably, knowledge of the channel characteristics is vital [13]. Key channel characteristics that influence selection of signaling parameters include delay dispersion, frequency coherence, Doppler spread, and temporal correlation. Knowledge of these characteristics enables optimal selection of transmission parameters (e.g., subcarrier bandwidth, symbol rate), as well as design parameters for remedial measures to counteract channel effects (e.g., equalization, diversity) [14]. Public safety communication systems in the 700 MHz and 4900 MHz bands are yet to be widely deployed. Consequently, characterization of wireless channels in these bands for emergency-responder environments is presently needed. The Department of Justice Community-Oriented Policing Services (COPS) program has funded NIST's Public Safety Communications Research Laboratory (PSRL) for several efforts in this area, including work described in [15]-[18]. The focus of NIST's work is dissemination in the open literature of measured wireless-channel characteristics in representative public safety environments using methods that can be reproduced by other researchers. Characterization of the urban channel, the subject of this technical note, represents a continuation of this work.

The work described here fills a gap in the existing literature by presenting channel measurement and modeling results—including spatial channel characteristics and the distribution of the number and delays of multipath components—for ground-based, peer-to-peer urban channels in the 700 MHz and 4900 MHz public safety frequency bands. These data are not currently available for human-height antennas. The measurement techniques we used (described in [11]) provide complex, densely sampled power delay profiles that can be used to quantify multipath clustering [19], develop new channel models [20], and verify more realistic, laboratory-based measurement methods such as those based on the reverberation chamber [21].

The remainder of this technical note is organized as follows: Section 2 describes the signal measurement process. Section 3 describes the urban environment and measurement parameters.

Section 4 presents delay dispersion characteristics for both LOS and non-LOS conditions. We also describe a path-loss model based on our data. Section 5 provides conclusions.

2. Measurement and Channel Characterization Methods

2.1. Wideband Channel Response Measurements

We measured the wideband frequency response and, from these, derived the time-delay characteristics of the propagation channel. We used a measurement system based on a vector network analyzer (VNA), shown in Figure 1. In other data collections [8], [18], and [22], we used this instrument to collect data over a very wide frequency range, from 25 MHz to 18 GHz. For the measurements reported here, we captured data from two frequency bands 700 MHz to 800 MHz, and 4.9 GHz to 5.0 GHz. This system, also described in [18], [22] lets us measure the complex transfer function of the channel, including frequency-selective characteristics. By taking the Fourier transform of the measured transfer function, the power delay profile and root-mean-square (RMS) delay spread of the channel are found in post processing.

The VNA acts as both transmitter and receiver in this system. The transmitting section of the VNA steps through the frequencies a single frequency at a time. The signal is amplified and fed to a transmit antenna, as shown in Figure 1. The received signal is returned to the VNA via a fiber optic cable. Sending the received signal along the fiber optic cable back to the VNA eliminates the loss and phase changes that would be associated with RF coaxial cables between the receive antenna and the transmit antenna, allowing characterization of the complex radio channel. One advantage of this system is that it provides a high dynamic range when compared to true time-domain-based measurement instruments. One disadvantage is that a time-varying channel may change during the long acquisition time.

In Figure 1, the system is configured for a LOS reference measurement. In practice, the transmit and receive antennas may be separated by significant distances, although they must remain tethered together by the fiber optic link. Omnidirectional antennas were used in our measurements because they are most often used in public safety applications. We used a discone for 700 MHz, and a monopole for 4900 MHz, both vertically polarized. Identical antennas were used at transmit and receive sites. Our four-port VNA enabled us to connect simultaneously to separate antennas for the 700 MHz and 4900 MHz bands. The synchronized fiber optic link between receive antennas and the VNA receive port enabled us to attain link distances up to 200 m.

In this work, we report path loss relative to a 4 m reference. The transmit antennas were set on a cart, and mounted to a positioner. The positioner is a motor-controlled, two-arm device that enabled us to move the transmit antennas in a Cartesian coordinate plane parallel to the ground. The positioner range is 0.5 m by 0.5 m. The receive antennas were mounted on tripods, and were moved manually from location to location. All antenna heights were approximately 1.6 m above the ground (to the top of the antennas).

To make a measurement, the vector network analyzer is first calibrated by use of standard techniques where known impedance standards are measured. The calibration enables us to correct for the response of the fiber optic system, amplifiers, and any other passive elements and electronics used in the measurement. We also high-pass filter our measurements in post processing to suppress the large, low-frequency oscillation that occurs in the optical fiber link. Because the received signals measured during our field tests tend to be quite weak, an amplifier is used. Consequently, during calibration of the VNA, an attenuator is inserted in the "thru" calibration standard path. This extra attenuation is corrected for in post processing for path-loss measurements.

For the measurements reported here, the VNA-based measurement system was set up with the following parameters: the initial output power was set to approximately 14 dBm. The gain of the amplifier and the optical link and the system losses resulted in a received power level no more than 0 dBm. An intermediate-frequency (IF) averaging bandwidth of around 1 kHz was used to average the received signal. We typically recorded 101 points per frequency band, with a frequency spacing of 1 MHz. The total sweep time was approximately 2 ms, giving a dwell time of approximately 20 μs per point. For the calibration process, the external attenuation was 40 dB for the high bands and 20 dB for the low bands.

2.2. Path Loss and Delay-Spread Calculations

2.2.1 Path loss

Our wideband measurements provide a channel transfer function $H(f)$, where $H(f)$ typically is derived from the measured transmission parameter $S_{21}(f)$. Because S parameters are ratios of voltage traveling waves that cannot provide absolute power, we find path loss with respect to a free-space reference measurement. To find the frequency-dependent path loss between the transmit and receive antennas, we first compute $|H(f)|^2/|H_r(f)|^2$, where $H_r(f)$ is a free-space reference made at a known distance d_r from the transmit antenna. The use of a ratio to find the path loss enables us to calibrate out the antenna response of the system. We correct the measurements for the free-space path loss between the transmit antenna and the reference location by dividing $|H_r(f)|^2$ by $(4\pi d_r/\lambda)^2$. The correction factor, i.e., $(4\pi d_r/\lambda)^2$, comes from the Friis free space equation:

$$P_r(d) = \frac{P_t G_t G_r}{L} \cdot \left(\frac{\lambda}{4\pi d}\right)^2, \tag{1}$$

where $P_r(d)$ is the received power at separation distance d (in meters), G_t and G_r are the gain of the transmit and receive antennas, L is the system loss factor not related to propagation, and λ is the free space wavelength. Note that $d = d_r$ in (1) when correcting for the reference location.

In addition, we can also compute the excess path loss, which is important for our discussion on measurement uncertainties to follow. To find the excess path loss, we additionally reduce the total path loss by the expected free-space path loss over the separation distance d between transmit and receive antennas. To do this, we divide the measurement of $|H(f)|^2$ by $(4\pi d/\lambda)^2$. Equivalently, we can multiply $|H(f)|^2/|H_r(f)|^2$ by $(d_R/d)^2$. The distance d may be measured or estimated from maps, depending on the environment. This provides the loss in excess of that which would be measured at the same distance in free space. We note that communication engineers typically think of excess path loss as a single-frequency or narrowband measurement. However, VNA measurements provide a much richer data set because they include both magnitude and phase information over a broad frequency range.

The reference measurement is made at a specified distance and may be acquired either during field tests or from a laboratory measurement. In the field, the measurement includes environmental effects, and we use time-domain gating to minimize reflections on the free-space reference, in other words, to isolate the LOS path form environment reflections. If we are not able to gate out the reflections satisfactorily, the reference measurement is made in a laboratory facility such as an anechoic chamber or an open-area test site. In this case, we use the same antennas and measurement system set-up as were used in the field. For the measurements shown below, we used a four-meter reference made at the NIST open area test site. The two antennas were mounted 1.6 m and 5 m above the ground plane, for the 4900 MHz and 700 MHz bands, respectively (see Figure 4). We chose this set-up to balance the need to be in the antenna far field of our lowest frequency of interest (700 MHz, with λ = 0.43 m in free space) for the results reported here, while keeping the reference measurement as free from environmental reflections as possible.

2.2.2 Delay Spread Data Processing

The time-domain representation of the signal was calculated from the path-loss data in post processing. Baseband (complex envelope) channel impulse responses were computed from the transfer functions by first windowing with a Hamming window to reduce delay-domain sidelobes. This technique is often employed with VNA measurements, for example [23]. Then the windowed transfer functions were inverse-Fourier-transformed to obtain channel impulse responses. For a channel impulse response denoted $h(\tau, t_i)$, the corresponding i^{th} ("instantaneous") power delay profile (PDP) was computed as $P_i(\tau) = |h(\tau, t_i)|^2$. The channel impulse response $h(\tau, t_i)$ represents the channel output at time t_i due to an impulse input at time $t_i - \tau$, and is given by [13]

$$h(\tau, t_i) = \sum_{k=1}^{L_{pi}} \alpha_{ki} e^{j\phi_{ki}} \delta(\tau - \tau_{ki}), \qquad (2)$$

where i indexes the ith PDP, L_{pi} is the number of multipath components in the ith PDP, and the amplitude and phase of the kth multipath component in the ith PDP are α_{ki} and ϕ_{ki}, respectively. The δ is a Dirac delta function, and τ_{ki} represents the delay of the kth multipath components of the ith PDP. Generally, α_{ki} and ϕ_{ki} are functions of time, but for each PDP in our case they can be considered constants due to the static nature of the scattering environment. Thus, the PDP $P_i(\tau)$ can be expressed as

$$P_i(\tau) = \sum_{k=1}^{L_i} |\alpha_{ki}|^2 \, \delta(\tau - \tau_{ki}).$$ (3)

In order to separate actual multipath components from noise, we also gathered pure noise transfer functions, denoted by $N(f)$. These were obtained with the VNA transmit ports terminated in matched loads, so the receive antennas received "ambient" noise only. From the $N(f)$ transfer functions, we computed complex baseband time-delay-domain noise samples. The noise samples were judged Gaussian by computing the Kullback-Leibler divergence D_{KL} [24], a goodness-of-fit measure for probability density functions. If D_{KL} equals zero, the fit to the Gaussian density is perfect; as D_{KL} increases, the goodness of the fit decreases. Values of D_{KL} were always less than 0.06, and typically $D_{KL} < 0.02$, indicating that the noise can be considered Gaussian.

We then set a noise threshold by means of the algorithm in [25], based upon the measured noise variance. All PDP samples below the noise threshold were discarded, so that the false-alarm probability means that for each sample in $P_i(\tau)$, the probability of mistaking a noise component for a multipath component was 0.01, or one noise sample mistaken for a multipath component per 100 PDP samples. Figure 5 (a) shows typical PDPs for LOS and NLOS conditions for the 700 MHz band, and Figure 5 (b) is an analogous figure for the 4900 MHz band. Note that all PDPs were normalized and delay-aligned by time-shifting back by the estimated direct-path delay, which was based upon the measured distances in Table I. In addition, we truncated all PDPs after collecting the first 99 % of the PDP energy. In Figure 5, a dynamic range is also indicated. We define the dynamic range as the difference in decibels between the PDP peak and the noise threshold. Mean dynamic ranges were above 18 dB for NLOS cases, and above 39 dB for LOS cases. Several PDPs had dynamic range less than 10 dB, and these were judged as having too low a signal-to-noise ratio, and were therefore discarded.

From the PDPs, we found the RMS delay spread. RMS delay spread is calculated as the second central moment of the power-delay profile of a measured signal [26]-[28]. Figure 3 shows the power-delay profile for a representative building propagation measurement. The peak level usually occurs when the signal first arrives at the receive antenna, although in high multipath environments we sometimes see the signal build up over time to a peak value and then fall off.

A common rule of thumb is to calculate the RMS delay spread from signals at least 10 dB above the noise floor of the measurement [28]. For the measurements described in the following sections, we defined the minimum dynamic range to be approximately 40 dB below the peak value, although this value was reduced for lower signal levels. For the illustrative measurement shown in Figure 3, we extended the window down to 70 dB below the peak value. Whether we use a 40 dB or a 70 dB threshold, the RMS delay spread does not change appreciably due to the almost constant slope of the power decay curve.

The RMS delay spread σ_τ can be defined as

$$\sigma_\tau = \sqrt{\overline{\tau^2} - \overline{\tau}^2}.$$ (4)

In (4), $\overline{\tau}$ is defined as the average value of the power-delay profile in the defined dynamic range window, and $\overline{\tau^2}$ is the variance of the power-delay profile within this window. Another delay spread measure that is sometimes used is the delay window [29]. The delay window is the duration that contains x % of the channel impulse response energy, and this is denoted $W_{\tau,x}$; Table 2 lists average PDP delay windows for $x = 90$ % energy. To determine $W_{\tau,x}$, we use a symmetric window that finds the "middle" x % energy. That is, for our example with $x = 90$, the delay window neglects the earliest 5 % and the latest 5 % of the channel impulse response energy. The last delay-domain dispersion measure that we list is the delay interval [29]. The delay interval $I_{\tau,X}$ is defined as the duration of the channel impulse response containing all impulses above X dB down from the largest impulse.

3. Experiment Set-up and Measurement Uncertainties in Denver, Colorado

3.1. Experiment Set-up

The measurements were taken outdoors in the financial district of downtown Denver on Saturday, June 20, 2009. This area contains many large (over 20-story) buildings. Figure 6 shows an illustration of the test area constructed from a Google map view.[1] The test area was in the block between 17th and 18th Streets, and between Welton Street and Glenarm Place. Figure 7 shows the 17th Street entrance into the building complex. Street widths are on the order of 20 m. Three transmitter locations and eleven receive antenna locations were used for a total of 33 transmit/receive location pairs. Figure 6 also shows a photograph of the two receiver antennas located at position R5 on the corner of Welton and 17th Streets. Figure 8 shows the sidewalk areas where the experiments took place, and Figure 9 shows the three transmitter locations. LOS link distances ranged from 10 m to 80 m. NLOS link distances are described in one of two ways:
   (1) by an "L-shape," with the first distance d_1 corresponding to the LOS distance from the transmit to a corner (e.g., T_1 to R_5 in Figure 6), and the second distance d_2 corresponding to the distance from the corner to the receive (e.g., R_5 to R_9 in Figure 6); or,
   (2) by a "U-shape," with d_1 as previously defined, d_2 the corner-to-corner distance, and d_3 defined as the final distance from the second corner to the receive (e.g., R_9 to R_{10} in Figure 6). Table 1 lists all these distances. Note that there is no Rx1 data; Rx1 was located indoors, and those data are not included in this paper. This approach for specifying distances in the urban environment has been used by others, for example [30].

For each band, we measured $H(f)$ twice at each of the nine transmit antenna positions (for each physical transmit/receive location pair), yielding 18 transfer functions per transmit/receive location pair. The nine transmit antenna positions corresponded to nine points on the Cartesian grid of the positioner, with separation between each point equal to 25 cm in both dimensions. Relative separation of grid positions is different in the two bands, but our results and analysis for each band individually are unaffected by this. Spatial channel information for both bands should be of interest even if array characteristics are not identical in the two bands.

---

[1] © 2009 Google, Map Data © 2009 Tele Atlas.

Figure 10 shows a plan view diagram of the positioner, with the nine individual antenna positions labeled P1-P9. With 33 transmit/receive pair combinations, a total of 18×33=594 transfer functions were collected. The measurement bandwidth for both bands was set to 100 MHz, with, but strong interference in the lower part of the 700 MHz band reduced that usable bandwidth to 75 MHz. Our transfer functions thus covered the 725 MHz to 800 MHz and 4.9 GHz to 5 GHz frequency bands. The frequency resolution of 1 MHz enabled a time measurement resolution of a 1μs maximum delay.

For much of the time during measurements, only pedestrian motion was present, although automobile traffic increased as the day progressed. Traffic around the block was "stop and go," since stoplights were present at all intersections. Auto speeds were as large as approximately 10 m/s, which for single-scattering yields a maximum Doppler frequency at 5 GHz of $f_d = vf/c = 10(5\times10^9)/3\times10^8 \cong 167$ Hz [13]. This yields a minimum channel coherence time of approximately $t_{c,min} \cong 1/f_d = 6$ ms for the 4900 MHz band. The same maximum velocity yields a minimum coherence time of approximately 37 ms for the upper end of the 700 MHz band. With each VNA sweep across the band taking approximately 2 ms, we assume the channel can be considered statistically wide sense stationary for each measured $H(f)$, especially since most vehicle velocities were less than our cited maximum. With our measurement procedure, we were unable to measure fading dynamics. Studies of the fading statistics of this propagation environment, including the Ricean K-factor, are currently a subject of research at NIST.

3.2. Measurement Uncertainties

Following the convention described in [31], the uncertainties associated with this measurement process can be broken into two categories, Type A (random) and Type B (systematic). The wireless channel is inherently non-static with respect to time, frequency, and position, and the data collected are impacted by these parameters. However, the measurement system itself does not change between data collections. Table 3 describes the various uncertainties associated with the measurement system while Table 4 contains additional uncertainties in our estimate of the mean path loss, such as the variability of the measured channel response, also known as small-scale fading (see [13],[14]).

To quantify the repeatability of the VNA measurements, which is a Type A uncertainty, we use multiple measurements made in the controlled environment of the NIST Open Area Test Site (OATS). This is a 30 m x 60 m ground plane located many electrical wavelengths from the nearest reflective objects or scatterers. We utilized a set of reference measurements conducted with the same antennas and measurement set-up used in Denver, with the exception of a long coaxial cable between the antenna and the VNA. Free-space measurements were collected at 2 m increments for antenna spacings from 4 m to 10 m for the 700 MHz band and at 1 m increments from 2 m to 5 m for the 4900 MHz band. Figure 11 shows the excess path loss for this series of reference measurements, calculated by selecting one of the measurements as the reference (the 0 dB result is the reference measurement). Four different sets of measurements were performed, two covering the 700 MHz band and two covering the 4900 MHz band. In each band, one set of measured data covered only the band of interest, that is, the 725 MHz to 800 MHz and the 4900 MHz to 5000 MHz frequency ranges. The third set of measurements was conducted between 300

MHz and 1 GHz. From this, data around the 700 MHz band were extracted. The fourth set of data covered 1 GHz to 10 GHz. From this, data around the 4900 MHz band were extracted. (Note that a 75 MHz range was used in the 700 MHz band rather than the full 100 MHz because in the field tests there was significant interference below 725 MHz, and, thus, the subsequent data processing only used the 725 MHz to 800 MHz data.) We corrected for differences in distance using a free-space assumption (see (1)) so that measurements made at each separation could be averaged directly. We then computed the standard deviation of the eight measurements.

For the 4900 MHz reference measurements, we computed a standard deviation of 0.55 dB for the eight measurements. The antenna heights for these measurements were 1.6 m, and therefore some ground reflections likely increased the standard deviation of these measurements. The separation between antennas was measured using a tape measure. This likely increased the standard deviation as well. For the 700 MHz band, the antennas were located at 5 m above the ground, resulting in less ground reflection contributions. The standard deviation for the eight measurements was approximately 0.3 dB. Thus, the estimated Type A uncertainty for the system, u_{sys}, would be 0.3 dB for the 700 MHz band and 0.55 dB for the 4900 MHz band.

We next consider the Type B uncertainties of the measurement system. The Type B uncertainties include the impact of temperature changes on the system measurement of S_{21}. The analysis in [32] estimated a VNA drift over three days in a laboratory setting as 0.1 dB, or $u_{\text{VNA}} = 0.1$ dB. S_{21} data collected on the fiber optic link over several days in a laboratory setting provides an estimated Type B uncertainty associated with the fiber link at 0.1 dB, i.e., $u_{\text{fiber}} = 0.1$ dB.

The combined uncertainty for a measurement of path loss using our VNA measurement set-up is then

$$u_{\text{measurement system,combined}} = \sqrt{u_{\text{sys}}^2 + u_{\text{fiber}}^2 + u_{\text{VNA}}^2}. \tag{5}$$

For the measurement set-up we used in Denver, this value is 0.33 dB and 0.57 dB for the 700 MHz and 4900 MHz bands, respectively.

A second source of random, Type A, uncertainty in our estimate of the path loss in the urban canyon can be attributed to small-scale fading arising from multiple reflections in the local area around each test location. Even though the building environment is deterministic, small-scale fading is considered random due to its extreme sensitivity to antenna placement and the fact that cars, truck, and pedestrians moved randomly through the environment during testing. By acquiring path loss data over multiple frequencies and positions for each transmit/receive antenna location, the effects of small-scale fading can be averaged out. Thus, experiment design, rather than equipment repeatability, leads to this source of uncertainty. Such variability in the channel is of interest to communications engineers and is reported separately here before being combined with the VNA measurement uncertainty.

The Type A standard uncertainty in our estimate of the mean path loss at each measurement

13

location is based on the standard deviation of the eighteen measurements (two measurements per each of the nine positioners locations) made at each of the twelve transmit/receive locations. We consider the behavior exhibited over a local region, i.e., the nine locations of the positioner, and calculate the standard deviation over frequency at each of the nine positions. We then calculate the standard deviation arising from small-scale fading between each of the nine positions. These two uncertainties, labeled u_{spacing} and u_{freq}, are provided in Table 4. Table 3 also lists the mean path loss, calculated over each of the nine positioner locations and over the 75 (700 MHz) or 100 (4900 MHz) frequencies. Because we made each measurement twice, 18 individual estimates of the path loss are averaged to obtain the value reported in Table 3.

The combined uncertainty in a mean path loss measurement is calculated as:

$$u_{\text{c}} = \sqrt{u_{\text{VNA}}^2 + u_{\text{fiber}}^2 + u_{\text{sys}}^2 + u_{\text{spacing}}^2 + u_{\text{freq}}^2}. \tag{6}$$

Table 4 lists the combined uncertainty values for all the transmit/receive antenna location pairings. The largest contribution to the uncertainty in the mean path loss is due to the standard deviation across the frequency band. The mean path loss value (based on a 4 m reference location) and error bars for ± two standard deviations at the twelve transmit/receive locations are shown in Figure 12 and Figure 13 for the 700 MHz and 4900 MHz bands, respectively. In this case, the standard deviation represents the standard uncertainty in the estimate of the mean path loss and the error bars represent a coverage factor of two. Under the assumption of a normal distribution in mean path loss values, the coverage factor of two corresponds to a 95% confidence interval.

4. Experiment Results

4.1. Path Loss Results

With a known reference path loss and a calibrated VNA, we also used our VNA measurements to estimate propagation path loss. For each transmit/receive location pair, we have 18 path loss measurements. These correspond to three groups of six measurements on our Cartesian positioner: group one represents the six measurements (two measurements at each of the three positioner points) at distance d, group two represents the six measurements at distance $d + 0.25$ m, and group three represents the six measurements at distance $d + 0.5$ m. (Recall that we have nine points on our Cartesian grid for each transmit antenna location, and at each point we collected two transfer functions—see Figure 3.)

To attempt to average out the effects of small-scale (multipath) fading, we estimated the path loss as the difference between the known transmit power at band center (equal to the power at any frequency in the band) and the average of the magnitude squared of the received transfer function, with the average taken across all frequency points in the band (75 MHz for the 700 MHz band, and 100 MHz for the 4900 MHz band). As discussed, measurements were made with respect to a 4 m reference; that is, antenna gains are also removed.

14

Figure 14 shows a plot of path loss in decibels versus $10\log_{10}(d/d_r)$ for the 700 MHz band, transmitter location T1, where d_r=4 m is the reference distance. This reference distance was also used for the 4900 MHz band. Reference measurements were made at the NIST open area test site in Boulder, CO which closely simulates free-space conditions, with the nearest reflecting objects well outside the time window of the measurement. Ground reflections were mitigated by elevating the antennas sufficiently during the reference measurement. Linear fits on the log-log scale are also shown in Figure 14. These are least-squares fits, and for the LOS case, these fits correspond to the following path loss model:

$$L_{\mathrm{LOS}}(d) = L(d_r) + 10n_{\mathrm{LOS}}\log(d/d_r) + X_{\mathrm{LOS}}, \qquad (7)$$

where n_{LOS} is the propagation path loss exponent, and X_{LOS} is a zero-mean Gaussian random variable with standard deviation σ_X dB. For NLOS L-shaped paths we used

$$L_{\mathrm{NLOS}}(d_1, d_2) = \tilde{L}_{\mathrm{LOS}}(d_1) + 10n_{\mathrm{NLOS}}\log(d_2/d_1) + L_c + X_{\mathrm{NLOS}}, \qquad (8)$$

where n_{NLOS} and X_{NLOS} are analogous to the LOS parameter definitions, distances d_1 and d_2 correspond to L-shaped path distances (see Section 3), and $\tilde{L}_{\mathrm{LOS}}(d_1) = L_{\mathrm{LOS}}(d_1) - X_{\mathrm{LOS}}$, so that we do not apply the Gaussian random variable twice for the NLOS path loss. The parameter L_c is the "corner loss" added to the LOS path loss at distance d_1; this loss term was also used in [16]. The corner loss corresponds to the "step" discontinuity between the LOS and NLOS fits; see Figure 14. The intercept value $L(d_r)$ is equal to the free-space value at band center for both bands; i.e., $L_{700\mathrm{MHz}}(d_r)$=42 dB, and $L_{4900\mathrm{MHz}}(d_r)$=58 dB.

Table 5 shows values for the path-loss model parameters. Path-loss exponents for the LOS case are less than those for NLOS regions, as expected. Based upon our transmit locations, the NLOS exponents also increase with the distance from the transmitter to the corner (d1 in Table 1). This same dependence on "corner distance" was also observed in [30]. Our measurements are in the smaller range of corner distances covered in [30], but the range of path-loss exponents found generally agrees with values given in [30]. For the LOS case, exponents less than two may indicate waveguiding by the urban canyon walls; this is most noticeable for the 4900 MHz band. For the 700 MHz band LOS results, fitted path-loss exponents are generally slightly larger than that for free space (n=2), except for transmit location 3; as per the model, path loss variation is quantified by the Gaussian random variables (X) in (7) and (8). Transmit location 3 was very close to the building wall on 17th Street, and was partly shadowed by several pillars that extended out from the wall. The NLOS exponents ranged from 3.6 to nearly 6.

4.2. Delay Spread Results

4.2.1 Power Delay Profile Results for Individual Transmit/Receive Antenna Pairings

As discussed in Section 3, for each individual pairing between a receive and a transmit antenna, we calculate the channel transfer functions. We measured a transfer function for each of the nine positions available with the positioner, and repeated the process twice (18 per transmit/receive pairing). Channel response $|H(f)|^2$ and PDP results are plotted in Appendix II (700 MHz) and

Appendix III (4900 MHz). The left-hand plots show the mean of $|H(f)|^2$ along with the standard deviation calculated over the nine positions and two repeat measurements. The mean of $|N(f)|^2$ is also shown. The right-hand plots show the corresponding PDP, with the mean and standard deviation of the τ_{rms} values calculated over position for the two repeats.

In Figure 15, the mean 700 MHz PDP results (calculated over nine positions and two repeats) are plotted versus receive antenna location for the three transmit locations, respectively. Similarly, Figure 16 plots the mean 4900 MHz PDP results against the receive antenna locations for the three transmit locations, respectively. These plots also include a plot of the standard deviation ($\pm\sigma$) to indicate the variably of the PDP at those locations. Note that data were not available for the Rx10-Tx3 calculation due to the insufficient margin between the signal and the noise to calculate the RMS delay spread.

For the 700 MHz results, Rx6, which is an NLOS pairing for all three transmitter locations, shows the highest mean RMS delay spread (over nine positions and two repeats) for Tx1 and Tx2 pairings, and the second highest mean for Tx3. The location of Tx3 is quite near a building surface and, thus, some difference in behavior compared to Tx1 and Tx2 is expected. Another interesting observation is the behavior of Rx10 in the 4900 MHz band, where the long RMS delay spread indicates that the signal propagates via multiple reflections (i.e., significant attenuation and/or multipath contribution) through the atrium of the building shown in Figure 7, even though a potential path of propagation exists through glass doors and an open lobby. This is readily observed by individual Rx9 and Rx10 results in Appendix II and III. In both the 700 MHz and 4900 MHz bands, both the $|H(f)|^2$ and corresponding PDP curves show the Rx10 values are much closer to the noise floor than the Rx9 values, and in some instances, the Rx10 values are below the noise floor. The PDP's in these cases are unlikely to be highly accurate.

4.2.2 Aggregate Power Delay Profile Results

The average power delay profiles, computed separately over all LOS PDPs and all NLOS PDPs and for each frequency band, are shown in Figure 17. Here, we see that the average PDPs look similar for the two bands. Table 2 provides delay-spread statistics for both cases and bands.

As expected, NLOS delay spreads are substantially larger than those for LOS cases. Also as expected, delay spreads generally increase with link distance [33]. The 4900 MHz band delay spread values are also slightly larger than those for the 700 MHz band. This relationship does not always hold; delay spreads generally (but not always) decreased with increasing frequency in [34], [35], but it is not clear whether signal-to-noise ratio and dynamic range were equal in all bands in the results of [34] and [35]. This means that comparison of delay spread trends may not be completely fair. This holds true in our case as well: because we used a higher-power external amplifier at the transmit end of the VNA for the 4900 MHz band, the 4900 MHz dynamic ranges were generally larger than those for the 700 MHz band; mean values of dynamic range were 38 dB for the 4900 MHz band and 28 dB for the 700 MHz band. This can account for some of the larger delay spreads we observed at 4900 MHz. In addition, results in [34] and [35] were not for ground-based settings.

Because our measurements spanned several hours, the propagation conditions did not remain constant; conditions also of course change with transmit-receive locations. This can be quantified in the delay domain by use of "instantaneous" delay spread measures [36]. Essentially, we compute the delay spread measures for *each* PDP individually. We can then collect statistics on these values over the sets of PDPs to quantify the range of variation of the delay spread measures. Figure 18 and Figure 19 show histograms of instantaneous RMS-DS for the two bands, over all transmit/receive locations. These plots demonstrate the expected result that the majority of the NLOS delay spreads are larger than the majority of the LOS delay spreads, and that the range of delay spreads is a significant fraction of the mean. We can quantify this range via the coefficient of variation $CV = \sigma_{RMS-DS}/\mu_{RMS-DS}$, the ratio of delay spread standard deviation to delay spread mean: values of CV here are 0.27 to 0.34 for NLOS and 0.49 to 0.56 for LOS cases. Table 6 provides additional statistics on the instantaneous RMS-DS, and Table 7 shows analogous statistics for the 90 % energy delay window $W_{\tau,90}$ and 25 dB delay interval $I_{\tau,25}$. As expected for this peer-to-peer, short-range setting, RMS-DS values are much smaller than those for cellular; for example, the COST207 typical urban cellular channel has RMS-DS ~ 1µs [34]. Our delay spread values are also substantially smaller than those in [34], [35] in which median delay spreads for their (elevated-antenna) measurements range from 300 ns to 700 ns in frequency bands from 430 MHz to 6 GHz.

4.2.3 Estimated Multipath Component Contribution

When creating channel models, one would like to know the number of multipath components present. We used the algorithm in [37] to estimate the number of multipath components, denoted L_p, in each transfer function. This algorithm, which is a modified multiple signal classification (MUSIC) algorithm for frequency estimation, uses the minimum description length criterion [38] to determine the number of multipath components, based upon modeling the time-invariant transfer function as a harmonic function of delay; that is, the Fourier transform of the ith channel impulse response is viewed as a function of delay τ at our given set of measured frequency points $\{f_{ki}\}$. Hence, from (2) and [38, (3), (4)], we have

$$H(f,t_i) = \sum_{k=1}^{L_{pi}} \alpha_{ki} e^{j\phi_{ki}} \exp(-j2\pi f \tau_{ki})$$

$$\rightarrow H(\tau,t_i) = \sum_{k=1}^{L_{pi}} \alpha_{ki} e^{j\phi_{ki}} \exp(-j2\pi f_{ki}\tau) \tag{9}$$

and this representation enables use of MUSIC on this dual function for estimating the discrete delays. Note that, to our knowledge, this algorithm has previously been used only for indoor channels. Summary statistics for the number of multipath components are presented in Table 8. These statistics count the number of components within a 25 dB threshold of the peak component in each PDP, where we truncated each PDP before applying the MUSIC algorithm. We employed this threshold because most communication systems typically do not operate at signal-to-noise ratios much larger than this value, and hence models that retain only the largest components are common, e.g., those within a 20 dB threshold in [39]. Typically, one would expect the NLOS cases to have a substantially larger number of multipath components than the

17

LOS cases, but the NLOS numbers are only slightly larger here. We attribute this to the lower dynamic range of the NLOS PDPs.

The distribution of the number of multipath components L_p was found to be best fit by a modified uniform distribution. Specifically, let L_{pmin} equal the minimum value of L_p and L_{pmax} its maximum value (see Table 8). We denote the probability mass function for L_p by $Pr(L_p=m)$, with integer $m \in \{L_{pmin}, L_{pmin}+1, \dots L_{pmax}\}$. The mass function has weight p_0 at L_{pmax} and is uniform with weight equal to $(1-p_0)/(L_{pmax}-L_{pmin})$ from L_{pmin} to $L_{pmax}-1$, or

$$\Pr(L_p = m) = \begin{cases} \dfrac{1 - p_0}{L_{p\max} - L_{p\min}}, & m = L_{p\min}, L_{p\min} +1, \dots L_{p\max}-1 \\ p_0, & m = L_{p\max} \end{cases} \tag{10}$$

Values of p_0 are also listed in Table 8. As seen from Table 8, 18-21 multipath components suffice for the 100 MHz channel at 4900 MHz, and 11-14 multipath components suffice for the 75 MHz channel at 700 MHz. For construction of channel models for smaller values of bandwidths, multipath components can be combined, as in [40], [41].

The distribution of the powers in these multipath components was obtained by fitting to the average PDPs of Figure 17. For these cases, we employed the following models:

$$P_{\tau,\text{LOS}}(\tau) = c_0 \exp(-c_1 \tau) \tag{11}$$

$$P_{\tau,\text{NLOS}}(\tau) = \sum_{k=1}^{N_c} b_{k0} \exp[-b_{k1}(\tau - \tau_k)], \tag{12}$$

where $N_c=3$ in (12) is the number of clusters of multipath components for the NLOS case. The use of clusters is common in channel models for other settings as well; for example, the indoor setting in [42] and the outdoor macrocell setting in [39]. Our clusters were based upon visual inspection of the average PDPs. They are better defined for the 700 MHz band data, and we see that the 4900 MHz band clusters tend to overlap more substantially (see Figure 17). The model coefficients are given in Table 9, where the first cluster delay $\tau_1=0$.

For the delays of the multipath components within clusters, other researchers have employed randomly distributed delays; for example Poisson in [42], or for ultrawideband channels, uniformly distributed delays [43], or Weibull distributed delays [44], [45]. If we base the delay distribution upon the average PDPs, uniform distributions of delays fit the LOS cases in intervals [0, 500 ns) for the 700 MHz band and [0, 550 ns) for the 4900 MHz band. The average PDPs for the NLOS cases could also be fit with uniformly distributed delays. Yet better models for the delay distributions were derived by collecting statistics on delays over *all* the measured PDPs. The results of this were that the LOS cases were best fit by an exponential distribution of delays, and the NLOS cases were best fit by a Weibull distribution [46] of delays. Specifically, the multipath component delay probability density functions are as follows:

$$p_{\tau,\text{LOS}}(\tau) = \exp(-\tau/v)/v \qquad (13)$$

$$p_{\tau,\text{NLOS}}(\tau) = \frac{\beta}{a^{\beta}} \tau^{\beta-1} \exp\left[-\left(\frac{\tau}{a}\right)^{\beta}\right], \qquad (14)$$

where in the Weibull density of (14), β is the shape factor and a is the scale parameter. Parameter values for these multipath component delay probability density functions are given in Table 10.

Finally, note that in other models (see [39], [42]), for simplicity, the decay constants b_{kl} of (12) for the decays within each cluster are assumed identical. As with our results here though, this is not always true: some ultrawideband models described in [43] employ different values of decay constants per cluster. If desired for simplicity, one could of course select a single value of decay constant from our models as well.

5. Summary of Results, and Conclusion

Here we reported on channel measurements and models for urban peer-to-peer, ground-based channels in the 700 MHz and 4900 MHz public safety frequency bands. This configuration and band has not been previously studied. Non-mobile outdoor measurements for link distances up to approximately 100 m were made with a vector network analyzer and omnidirectional antennas at a height of 1.6 m.

The uncertainties in our data associated with the measurement equipment are estimated at less than 0.6 dB. This is much smaller than the uncertainty caused by small scale fading within the channel itself, typically considered as the variability within the channel. This channel variability can also be viewed as a measurement uncertainty, and we that those values range from 4.5 to 7.1 dB. Thus, calculations based on these measured data such as the PDPs are not impacted significantly by the measurement equipment uncertainties.

For propagation path loss, we found path-loss exponents to range from 1.3-4.4 for LOS cases, and from 3.6-5.8 for NLOS cases around corners. In agreement with results found by other researchers, delay dispersion statistics appear similar for the two bands. The 90th percentile values for root-mean-square delay spread range from approximately 100 ns for LOS cases at 700 MHz, to 170 ns for NLOS cases at 4900 MHz, with maximum values of delay spread near 300 ns. We employed a "dual MUSIC" algorithm to determine the number of multipath components, and found that for our measurement bandwidths of 75 MHz at 700 MHz, and 100 MHz at 4900 MHz, mean values of the number of multipath components are 11 and 17, respectively. Least-square fits for the powers and delays of the multipath components were also computed, yielding complete statistical delay domain channel models.

Disclaimer: Mention of any company names serves only for identification, and does not constitute or imply endorsement of such a company or of its products.

19

We thank members of the technical staff of the Electromagnetics Division 687, who developed the measurement set-up, Mike Francis, Perry Wilson, Mike Kelley and Dereck Orr for programmatic support. This work was funded by the NIST Public Safety Communications Research Laboratory within the NIST Office of Law Enforcement Standards of NIST, Dereck Orr, Program Manager.

6. References

[1] Statement of Requirements: Background on Public Safety Wireless Communications, The SAFECOM Program, Department of Homeland Security, Vol. 1, March 10, 2004.

[2] M. Worrell and A. MacFarlane, Phoenix Fire Department Radio System Safety Project, Phoenix Fire Dept. Final Report, Oct. 8, 2004. http://www.ci.phoenix.az.us/FIRE/radioreport.pdf

[3] 9/11 Commission Report, National Commission on Terrorist Attacks Upon the United States, 2004. http://govinfo.library.unt.edu/911/report/index.htm

[4] Final Report for September 11, 2001 New York World Trade Center terrorist attack, Wireless Emergency Response Team (WERT), http://www.wert-help.org/WERTv2_files/WERT%20FINAL%20REPORT.pdfOct. 2001.

[5] C.L. Holloway, G. Koepke, D. Camell, K.A. Remley, D.F. Williams, S. Schima, S. Canales, and D.T. Tamura, "Propagation and Detection of Radio Signals Before, During and After the Implosion of a Thirteen Story Apartment Building," NIST Technical Note 1540, Boulder, CO, May 2005.

[6] C.L. Holloway, G. Koepke, D. Camell, K.A. Remley, and D.F. Williams, "Radio Propagation Measurements During a Building Collapse: Applications for First Responders," Proc. Intl. Symp. Advanced Radio Tech., Boulder, CO, March 2005, pp. 61-63.

[7] C.L. Holloway, G. Koepke, D. Camell, K.A. Remley, D.F. Williams, S. Schima, and D.T. Tamura, "Propagation and Detection of Radio Signals Before, During and After the Implosion of a Large Sports Stadium (Veterans' Stadium in Philadelphia)," NIST Technical Note 1541, Boulder, CO, October 2005.

[8] K. A. Remley, G. Koepke, C.L. Holloway, C, Grosvenor, D. Camell, J. Ladbury, J. T. Johnk, D. Novotny, W. F. Young, G Hough, M. C. McKinley Y. Becquet, J Korsnes, "Measurements to Support Broadband Modulated-Signal Radio Transmissions for the Public safety Sector," NIST Technical Note 1546, Boulder CO, April 2008.

[9] T. L. Doumi, "Spectrum Considerations for Public Safety in the United States," IEEE Comm. Mag., vol. 44, no. 1, pp. 30-37, January 2006.

[10] K. Balachandran, K. C. Budka, T. P. Chu, T. L. Doumi, J. H. Kang, "Mobile Responder Communication Networks for Public Safety," IEEE Comm. Mag., vol. 44, no. 1, pp. 56-64, January 2006.

[11] TIA, "TIA-902.BAAB-A Wideband Air Interface Scalable Adaptive Modulation (SAM) Physical Layer Specification – Public Safety Wideband Data Standards Project – Digital Radio Technical Standards," Sept. 2003, www.tiaonline.org.

[12]   IEEE 802 standards web site, http://standards.ieee.org/getieee802/portfolio.html, July 2010.

[13]   J. D. Parsons, The Mobile Radio Propagation Channel, 2nd ed., John Wiley & Sons, New York, NY, 2000.

[14]   A. F. Molisch, Wireless Communications, John Wiley & Sons, New York, NY, 2005.

[15]   C. L. Holloway, G. Koepke, D. Camell, K. A. Remley, S. A. Schima, M. McKinley, R. T. Johnk, "Propagation and Detection of Radio Signals Before, During, and After the Implosion of a Large Convention Center," NIST Technical Note 1542, June 2006.

[16]   C. L. Holloway, W. F. Young, G. Koepke, K. A. Remley, D. Camell, Y. Becquet," Attenuation of Radio Wave Signals Coupled into Twelve Large Building Structures," NIST Technical Note 1545, February 2008.

[17]   W. F. Young, C. L. Holloway, G. Koepke, D. Camell, Y. Becquet, K. A. Remley, "Radio Wave Signal Propagation Into Large Building Structures Part 1: CW Signal Attenuation and Variability," IEEE Trans. Ant. Prop., pp. 1279-1289, April 2010.

[18]   K. A. Remley, G. Koepke, C. L. Holloway, C. Grosvenor, D. Camell, J. Ladbury, R. T. Johnk, W. F. Young, "Radio Wave Signal Propagation Into Large Building Structures Part 2: Characterization of Multipath," IEEE Trans. Ant. Prop., pp. 1290-1301, April 2010.

[19]   H. Fielitz, K. A. Remley, C. L. Holloway, Q. Zhang; Q. Wu, D. W. Matolak, "Reverberation-Chamber Test Environment for Outdoor Urban Wireless Propagation Studies," IEEE Ant. Wireless Prop. Lett., pp. 52-56, March 2010.

[20]   C. Gentile, N. Golmie, K. A. Remley, C. L. Holloway, W. F. Young, "A channel propagation model for the 700 MHz band," Proc. IEEE Int. Conf. Comm (ICC 2010), May, 2010, pp. 1-6.

[21]   E. Genender, C. L. Holloway, K. A. Remley, J. M. Ladbury, G. Koepke, H. Garbe, "Simulating the Multipath Channel With a Reverberation Chamber: Application to Bit Error Rate Measurements," IEEE Trans. EMC, vol. 52, no. 4, 2010, pp. 766-777.

[22]   B. Davis, C. Grosvenor, R. T. Johnk, D. Novotny, J. Baker-Jarvis, M. Janezic, "Complex Permittivity of Planar Building Materials Measured with an Ultra-wideband Free-field Antenna Measurement System," Natl. Inst. Stand. Technol. J. Res., vol. 112, no. 1, Jan.-Feb., 2007, pp. 67-73.

[23]   A. Chehri, P. Fortier, P. M. Tardif, "Frequency Domain Analysis of UWB Channel Propagation in Underground Mines," Proc. IEEE Fall Veh. Tech. Conf., Montreal, CA, 25-28 September 2006.

[24] T. M. Cover, J. Thomas, Elements of Information Theory, 2nd ed., John Wiley & Sons, New York, NY, 1991.

[25] E. S. Sousa, V. M. Jovanovic, C. Daigneault, "Delay Spread Measurements for the Digital Cellular Channel in Toronto," IEEE Trans. Veh. Tech., vol. 43, no. 4, pp. 837-847, Nov. 1994.

[26] J. C.-I. Chuang, "The effects of time delay spread on portable radio communications channels with digital modulation," IEEE J. Selected Areas in Comm., vol. SAC-5, no. 5, June 1987, pp. 879-889.

[27] Y. Oda, R. Tsuchihashi, K. Tsuenekawa, M. Hata, "Measured path loss and multipath propagation characteristics in UHF and microwave frequency bands for urban mobile communications," Proc. 53rd IEEE Vehic. Technol. Conf., vol. 1, pp. 337-341, May 2001.

[28] J. A. Wepman, J. R. Hoffman, L. H. Loew, "Impulse response measurements in the 1850-1990 MHz band in large outdoor cells", NTIA Report 94-309, June 1994.

[29] International Telecommunications Union (ITU), "Multipath Propagation and Parameterization of its Characteristics," Rec. ITU-R P.1407.

[30] J. R. Hampton, N. M. Merheb, W. L. Lain, D. E. Paunil, R. M. Shuford, W. T. Kasch, "Urban Propagation Measurements for Ground Based Communication in the Military UHF Band," IEEE Trans. Ant. Prop., vol. 54, no. 2, pp. 644-654, February 2006.

[31] B. N. Taylor, C. E. Kuyatt, "Guidelines for Evaluating and Expressing the Uncertainty of NIST Measurement Results," NIST Technical Note 1297, September, 1994.

[32] J. B. Coder, J. M. Ladbury, C. L. Holloway, and K. A. Remley, "Examining the True Effectiveness of Loading a Reverberation Chamber: How to Get Your Chamber Consistently Loaded." IEEE International Symposium on Electromagnetic Compatibility, July 25-30, 2010.

[33] L. J. Greenstein, V. Erceg, Y. S. Yeh, M. V. Clark, "A New Path-Gain/Delay-Spread Propagation Model for Digital Cellular Channels," IEEE Trans. Comm., vol. 46, no. 2, pp. 477-485, May 1997.

[34] P. Papazian, "Basic Transmission Loss and Delay Spread Measurements for Frequencies Between 430 and 5750 MHz," IEEE Trans. Ant. Prop., vol. 53, no. 2, pp. 694-701, February 2005.

[35] R. J. C. Bultitude, T. C. W. Schenk, N. A. A. Op den Kamp, N. Adnani, "A Propagation-Measurement-Based Evaluation of Channel Characteristics and Models Pertinent to the

Expansion of Mobile Radio Systems to Frequencies Beyond 2 GHz," IEEE Trans. Veh. Tech., vol. 56, no. 2, pp. 382-388, March 2007.

[36]   A. F. Molisch, M. Steinbauer, "Condensed Parameters for Characterizing Wideband Mobile Radio Channels," Int. Journ. Wireless Inf. Networks, vol. 6, no. 3, pp. 133-154, 1999.

[37]   X. Li, K. Pahlavan, "Super-Resolution TOA Estimation with Diversity for Indoor Geolocation," IEEE Trans. Wireless Comm., vol. 3, no. 1, pp. 224-234, January 2004.

[38]   M. Wax, T. Kailath, "Detection of Signals by Information Theoretic Criteria," IEEE Trans. Acoust., Speech, Signal Processing, vol. ASSP-33, pp. 387-392, April 1985.

[39]   H. Asplund, A. A. Glazunov, A. F. Molisch, K. I. Pedersen, M. Steinbauer, "The COST 259 Directional Channel Model—Part II: Macrocells," IEEE Trans. Wireless Comm., vol. 5, no. 12, pp. 3434-3450, December 2006.

[40]   D. W. Matolak, I. Sen, W. Xiong, "The 5-GHz Airport Surface Area Channel—Part I: Measurement and Modeling Results for Large Airports," IEEE Trans. Veh. Tech., vol. 57, no. 4, pp. 2014-2026, July 2008.

[41]   3rd Generation Partnership Project (3GPP), "3GPP deployment aspects," Valbonne, France, Tech. Rep. TR 25.943 V5.1.0, Jun. 2002.

[42]   A. A. M. Saleh, R. A. Valenzuela, "A Statistical Model for Indoor Multipath Propagation," IEEE Journ. Selected Areas Comm., vol. SAC-5, no. 2, pp. 128-137, February 1987.

[43]   A. F. Molisch, "Ultrawideband Propagation Channels-Theory, Measurement, and Modeling," IEEE Trans. Veh. Tech., vol. 54, no. 5, pp. 1528-1545, September 2005.

[44]   R. Kattenbach, "Statistical Distribution of Path Interarrival Times in Indoor Environment," Proc. Spring IEEE Veh. Tech. Conf., Ottawa, Canada, pp. 548-551, 18-21 May 1998.

[45]   P. Yegani, C.D. McGillem, "A Statistical Model for the Factory Radio Channel," IEEE Trans. Comm., vol. 39, no. 10, pp. 1445-1454, Oct. 1991.

[46]   A. Papoulis and U. Pillai, Probability, Random Variables, and Stochastic Processes, 4th ed. New York: McGraw-Hill, 2001.

Appendix I: Experiment Setups and Locations

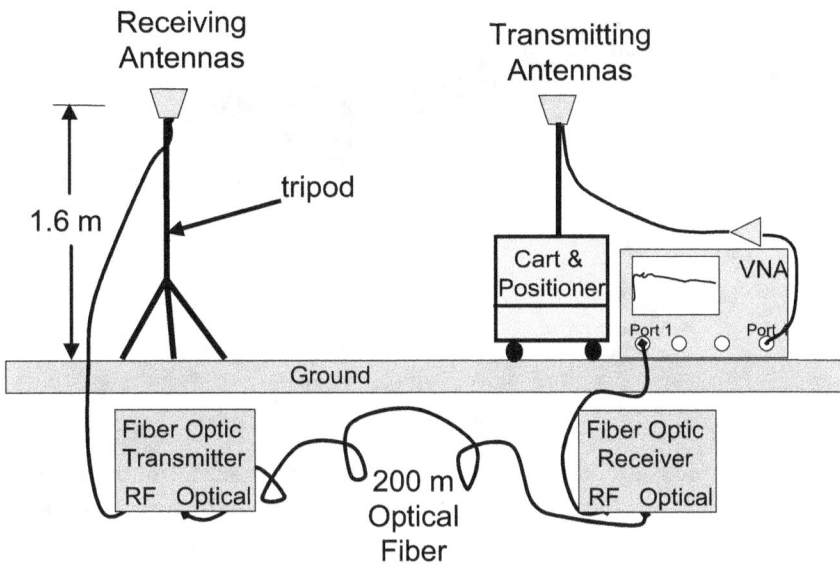

Figure 1. Wideband measurement system based on a vector network analyzer. Frequency-domain measurements, synchronized by the optical fiber link, are transformed to the time domain in post-processing. This enables determination of excess path loss, time-delay spread, and other figures of merit important in characterizing broadband modulated-signal transmissions.

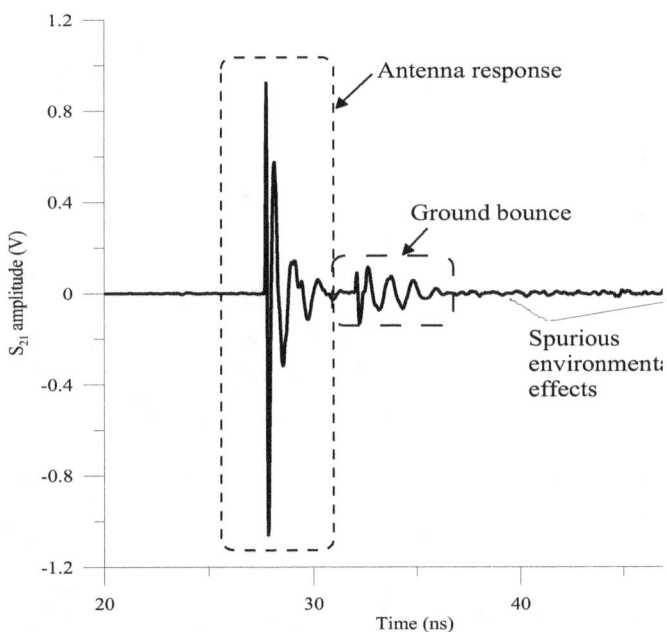

Figure 2. Reference measurement at three meters for a dual-ridge guide horn antenna, transformed to the time domain. The waveform shows the antenna response, the ground-bounce response and the spurious environmental effects.

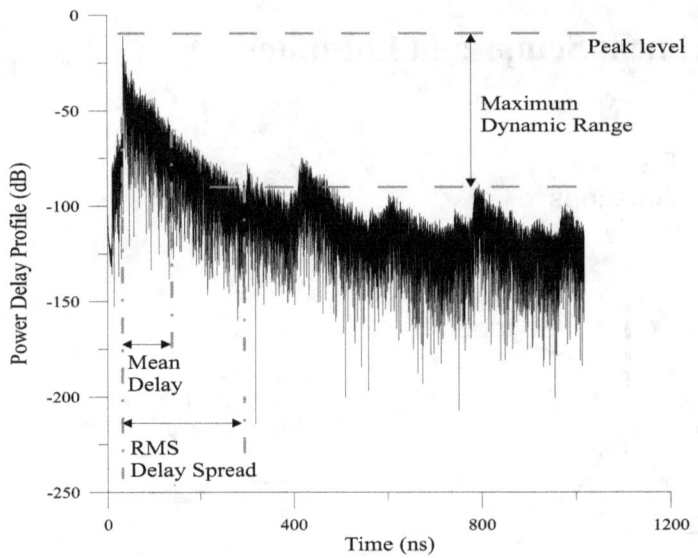

Figure 3. Power-delay profile for a building propagation measurement. Important parameters for a measured signal are the peak level, the maximum dynamic range, the mean delay, and the RMS delay spread.

Figure 4. Reference tests at the NIST, Boulder open area test site (OATS) with (a) 700 MHz and 4900 MHz antennas at a height of 1.6 m, and (b) 700 MHz antenna at a height of 5 m.

27

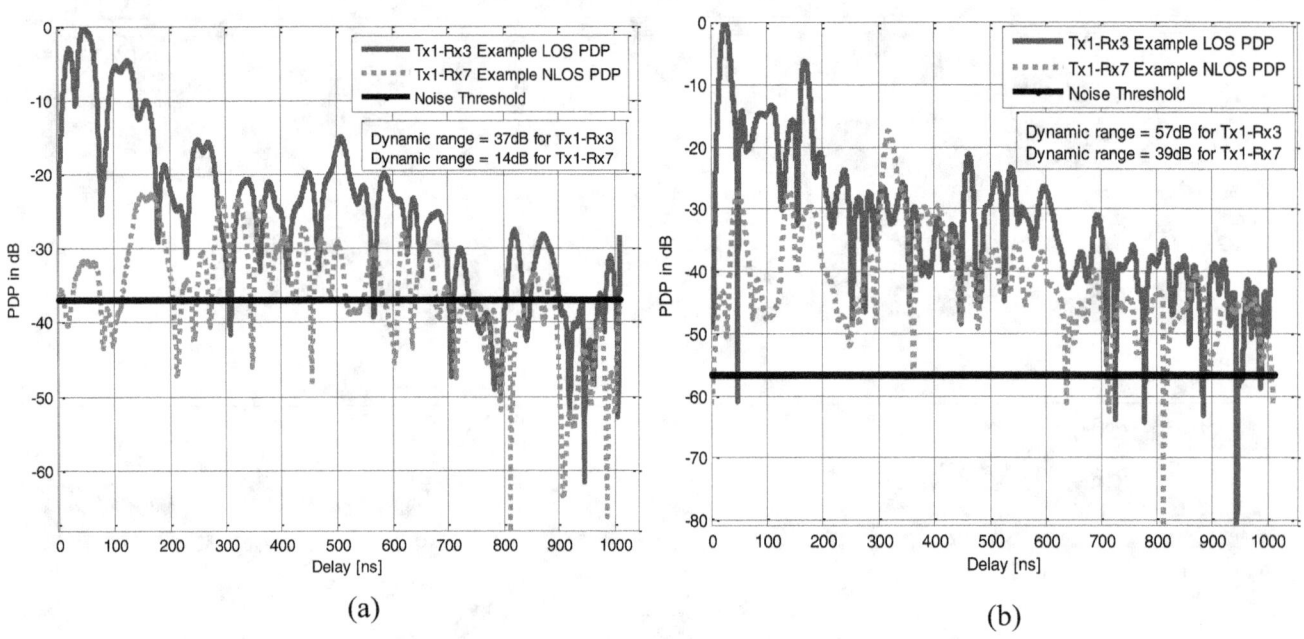

Figure 5. Example PDPs for (a) LOS (Tx1-Rx3) and NLOS (Tx1-Rx7) locations, 700 MHz band, and (b) LOS (Tx1-Rx3) and NLOS (Tx1-Rx7) locations, 4900 MHz band.

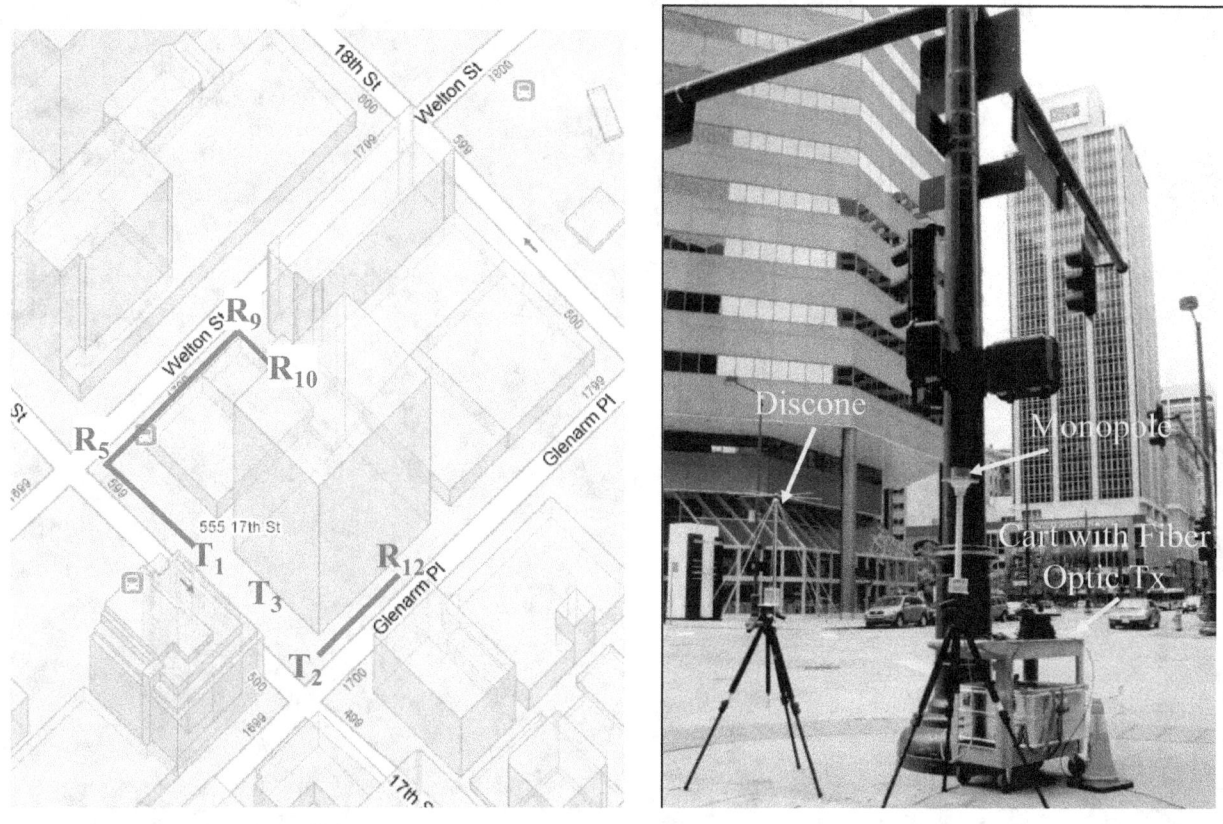

Figure 6. Google map view of test area in downtown Denver. Transmit locations denoted T, receiver locations denoted R. Right: photo of Rx antennas at location R5.

Figure 7. 17th Atrium at the 17th street entrance to the 555 17th Street Building.

Looking into the alley from Welton Street towards Glenarm Place (R_9 towards R_{10}).

Looking down the Welton Street sidewalk from towards 18th Street (R_5 towards R_9).

Looking down the 17th Street sidewalk from toward Welton Street (T_2 towards R_5).

Figure 8. Street level views of experiment location in Denver, CO at the 555 17th Street Building.

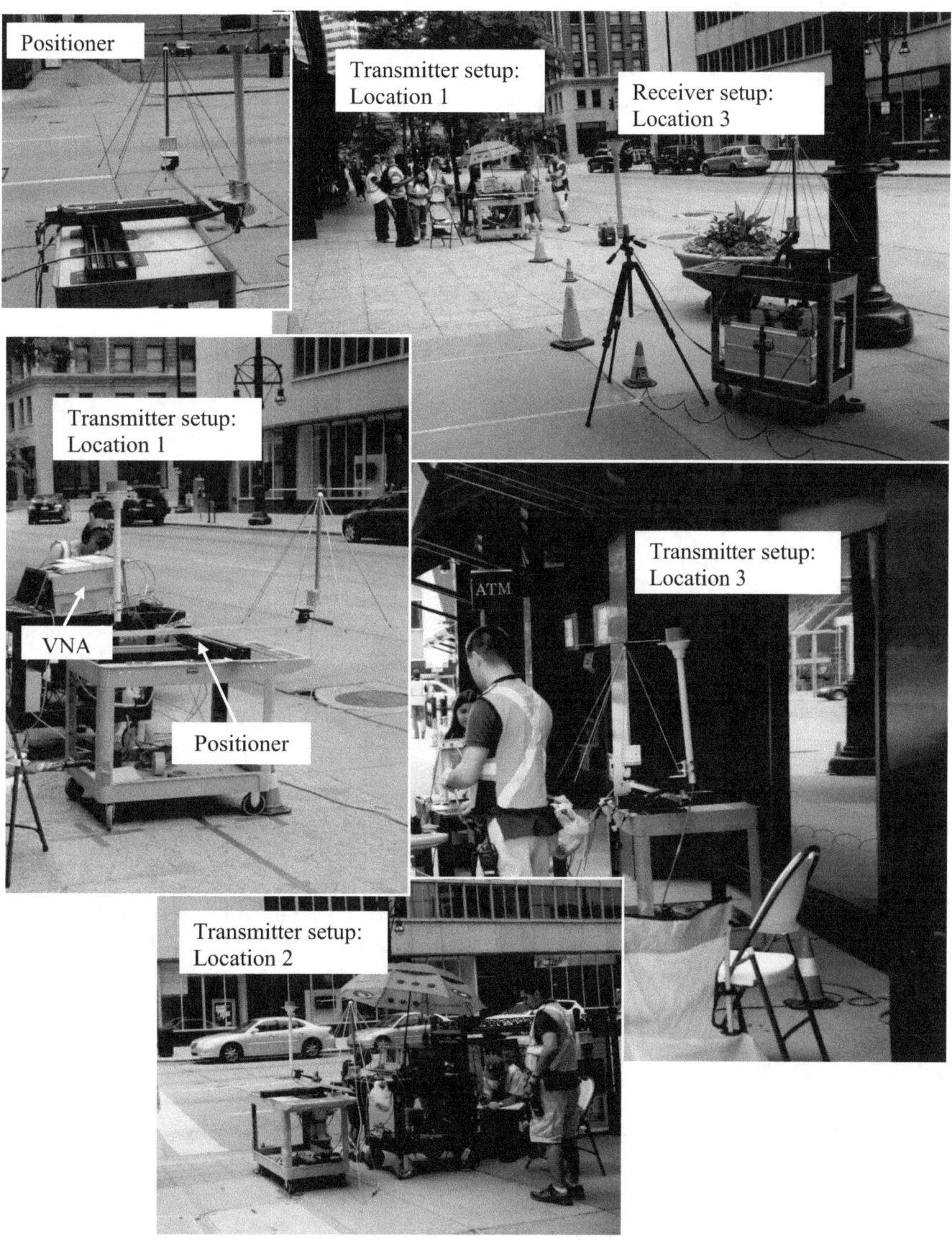

Figure 9. Transmitter sites for the experiments at Denver, CO.

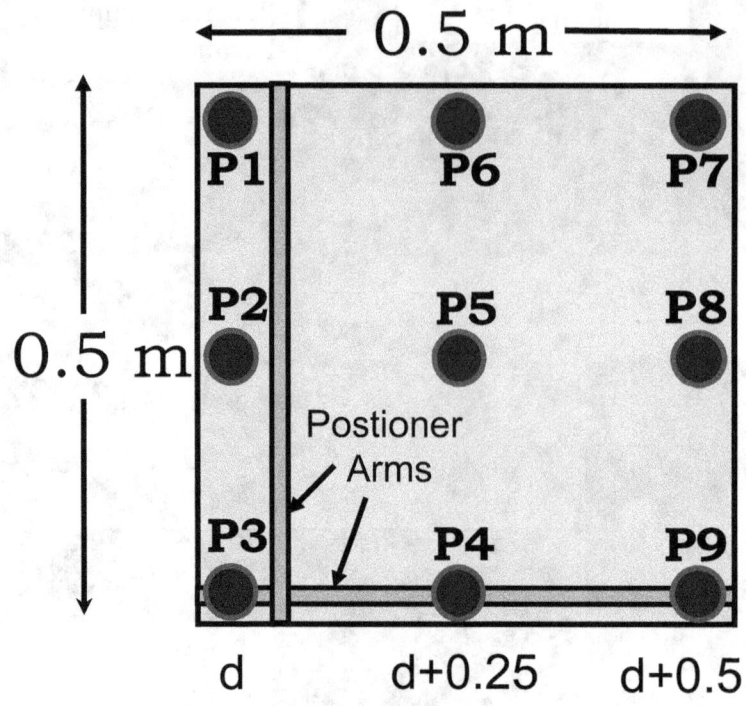

Figure 10. Plan view diagram of antenna positioner, showing nine positions.

Table 1. Transmitter (Tx) to Receiver (Rx) distances (m). LOS links contain only one distance (d), and NLOS links contain either two (d_1, d_2) for L-shaped paths, or three (d_1, d_2, d_3) for U-shaped paths.

	Rx2	Rx3	Rx4	Rx5	Rx6	Rx7	Rx8	Rx9	Rx10	Rx11	Rx12
Tx1	$d=10$	$d=20$	$d=30$	$d=40$	$d_1=40$ $d_2=10$	$d_1=40$ $d_2=20$	$d_1=40$ $d_2=30$	$d_1=40$ $d_2=40$	$d_1=40$ $d_2=43$ $d_3=13$	$d_1=40$ $d_2=5.5$	$d_1=40$ $d_2=35.5$
Tx2	$d=50$	$d=60$	$d=70$	$d=80$	$d_1=80$ $d_2=10$	$d_1=80$ $d_2=20$	$d_1=80$ $d_2=30$	$d_1=80$ $d_2=40$	$d_1=80$ $d_2=43$ $d_3=13$	$d=5.5$	$d=35.5$
Tx3	$d=36$	$d=46$	$d=56$	$d=66$	$d_1=66$ $d_2=10$	$d_1=66$ $d_2=20$	$d_1=66$ $d_2=30$	$d_1=66$ $d_2=40$	$d_1=66$ $d_2=43$ $d_3=13$	$d_1=14$ $d_2=5.5$	$d_1=14$ $d_2=35.5$

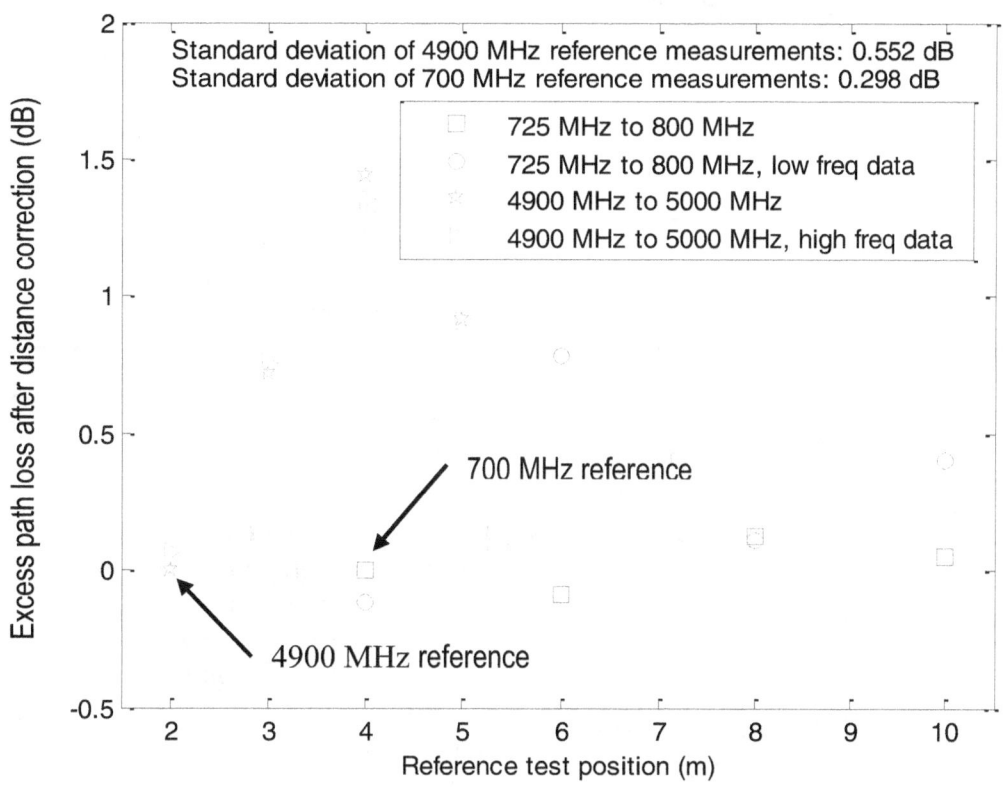

Figure 11. Excess path loss for the reference data collections performed at the NIST outdoor antenna test site. The excess path loss is calculated relative to the 700 MHz and 4900 MHz references, respectively, (i.e., the 0 dB cases), after correcting for separation distances assuming free space propagation.

Table 2. Summary Delay Spread Statistics (ns).

Condition (band)	RMS-DS	$W_{\tau,90}$	$I_{\tau,25}$
LOS (700)	66	166	386
LOS (4900)	87	235	519
NLOS (700)	147	501	798
NLOS (4900)	156	528	875

Table 3. List of the uncertainties in the data collection process.

Uncertainty Type	Uncertainty Description	Method of Estimate	Value (dB)
Type A 700 MHz band	Complete system through tests, including VNA, fiber optic link, and antennas	Standard deviation from eight independent system through measurements at Outdoor Antenna Site (OATS); corrected for free space to common distance of antenna separation	0.30
Type A 4900 MHz band	Complete system through tests, including VNA, fiber optic link, and antennas	Standard deviation from eight independent system through measurements at Outdoor Antenna Site (OATS) corrected for free space to common distance of antenna separation	0.55
Type A	Uncertainty in path loss estimation per position on the postitioner	Standard deviation of the path loss estimation across the frequency band, i.e., 725-800 MHz or 4900-5000 MHz	See "freq" column in **Table 4**
Type A	Uncertainty in path loss estimation between positions on the postitioner	Standard deviation of eighteen estimated means per location	See "spacing" column in **Table 4**
Type B	Drift in VNA measurements over time	Observed VNA drift over 3 days. See [32].	0.1
Type B	Impacts of temperature on fiber optic cable	Observation in controlled experiment over three days.	0.1

Table 4. Mean path loss and standard deviation for the eighteen measurements at each of the twelve locations. The combined column is the quadrature combination of the spacing, frequency band, system, fiber, and VNA uncertainties.

Location	TX 1 mean path loss	σ spacing	σ freq	σ combined	TX 2 mean path loss	σ spacing	σ freq	σ combined	TX 3 mean path loss	σ spacing	σ freq	σ combined
	700 MHz – all values in dB											
1	17.25	1.41	5.85	6.02	36.84	0.89	5.61	5.68	30.30	0.81	5.79	5.86
2	5.15	0.78	3.41	3.51	17.06	1.18	3.35	3.57	13.39	1.77	4.22	4.59
3	12.88	1.21	5.61	5.75	21.55	1.09	5.27	5.39	15.60	1.45	4.26	4.51
4	16.14	1.40	5.17	5.36	24.63	1.16	4.88	5.03	19.72	1.44	5.04	5.25
5	21.01	1.03	5.73	5.83	27.65	1.13	4.97	5.11	25.15	1.31	4.77	4.96
6	32.22	0.89	5.72	5.80	38.57	0.79	6.27	6.33	36.42	1.10	6.13	6.23
7	34.24	0.91	5.72	5.81	40.09	1.02	5.69	5.79	37.85	1.05	5.32	5.43
8	34.88	0.81	5.65	5.71	40.50	0.74	5.79	5.85	38.17	0.92	5.52	5.60
9	36.54	0.70	5.37	5.42	42.27	1.71	6.18	6.43	39.68	1.02	6.04	6.14
10	43.48	0.95	6.01	6.09	49.71	1.10	5.48	5.59	51.61	0.87	5.97	6.04
11	24.60	1.17	5.73	5.86	-1.26	1.27	2.41	2.74	18.14	1.40	5.78	5.96
12	36.42	1.03	5.21	5.32	22.03	1.11	5.84	5.96	32.97	0.88	5.98	6.06

Location	TX 1 mean path loss	σ spacing	σ freq	σ combined	TX 2 mean path loss	σ spacing	σ freq	σ combined	TX 3 mean path loss	σ spacing	σ freq	σ combined
	4900 MHz – all values in dB											
1	13.78	1.06	6.25	6.36	33.43	1.33	6.62	6.77	33.54	0.97	6.85	6.94
2	6.88	0.96	5.59	5.70	16.57	1.04	6.32	6.43	14.93	1.43	5.93	6.12
3	11.34	1.03	5.70	5.82	20.25	1.32	6.81	6.96	20.44	0.99	6.69	6.78
4	13.83	1.15	6.34	6.47	23.12	0.98	6.79	6.89	16.86	1.48	5.33	5.56
5	14.57	1.23	6.46	6.60	26.64	1.27	6.49	6.64	20.43	1.64	5.67	5.93
6	29.76	1.01	6.78	6.88	34.61	0.84	6.87	6.94	34.39	0.94	6.51	6.60
7	29.93	0.98	6.73	6.83	34.84	0.85	6.59	6.67	34.25	1.21	6.63	6.76
8	31.01	0.78	6.66	6.73	35.57	1.02	6.61	6.71	36.19	1.07	6.43	6.54
9	33.07	1.07	7.02	7.13	40.05	2.55	6.52	7.03	38.24	1.11	6.45	6.57
10	42.01	0.95	6.42	6.52	43.27	1.08	6.38	6.49	48.18	0.99	6.60	6.70
11	24.64	0.94	6.45	6.55	5.89	1.14	4.51	4.68	15.31	1.20	6.21	6.35
12	31.56	1.06	6.38	6.49	18.08	1.38	6.16	6.34	29.85	0.92	6.61	6.70

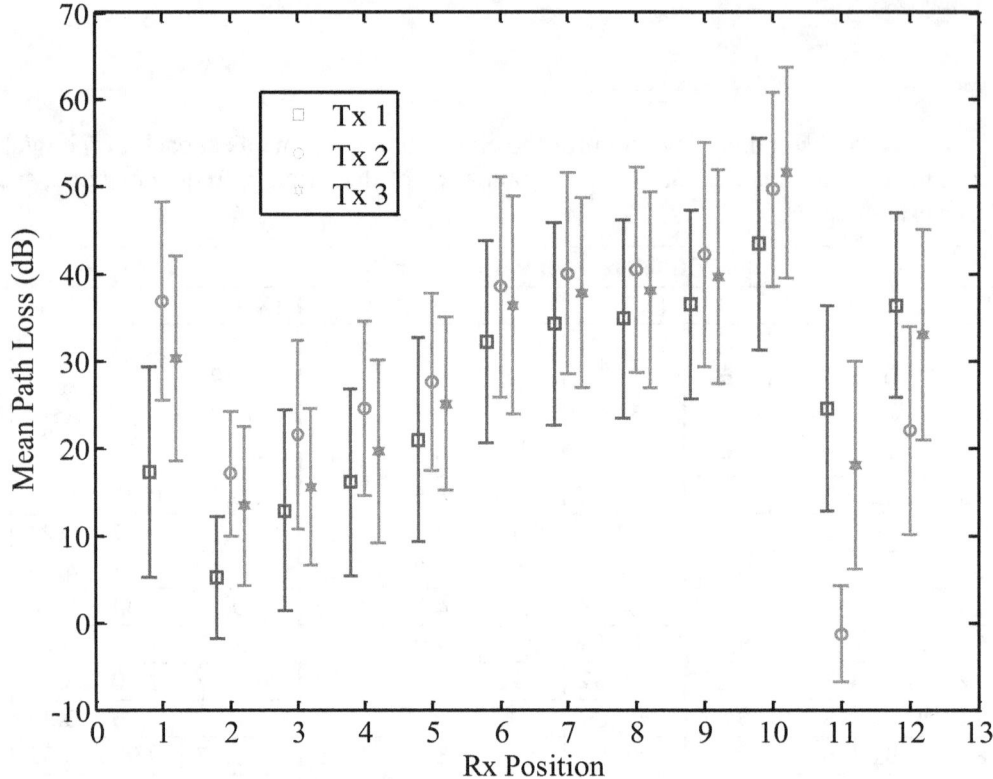

Figure 12. The mean path loss of the eighteen measurements collected at each of the twelve locations, with ± 2σ error bars, for each to the three transmitter locations. 725 MHz to 800 MHz.

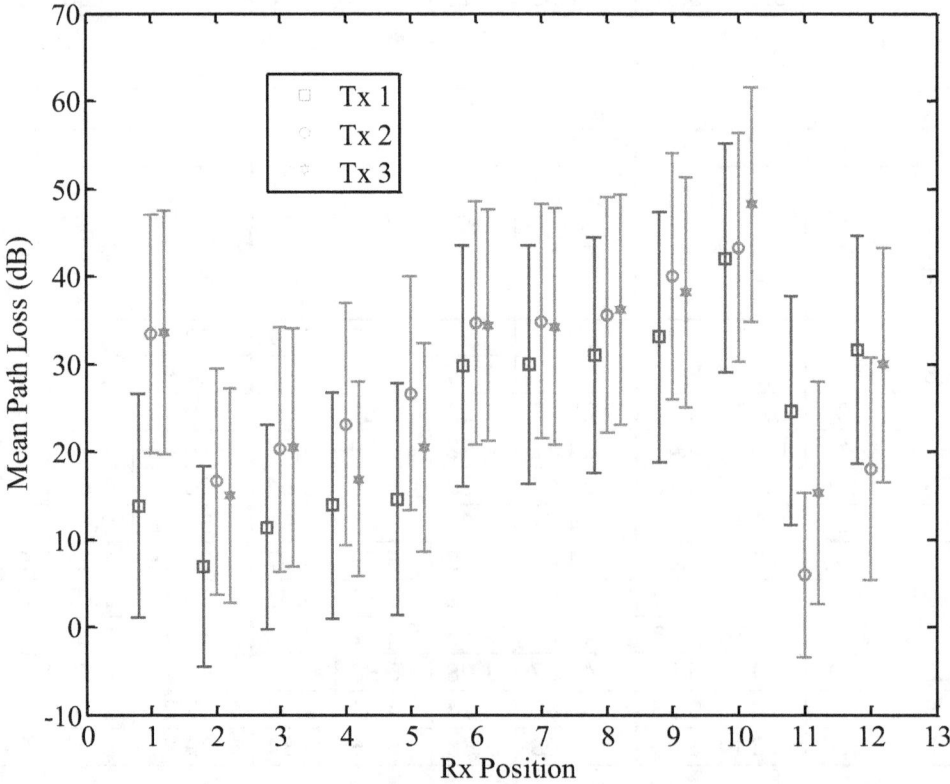

Figure 13. The mean path loss of the eighteen measurements collected at each of the twelve locations, with ± 2σ error bars, for each to the three transmitter locations. 4900 MHz to 5000 MHz.

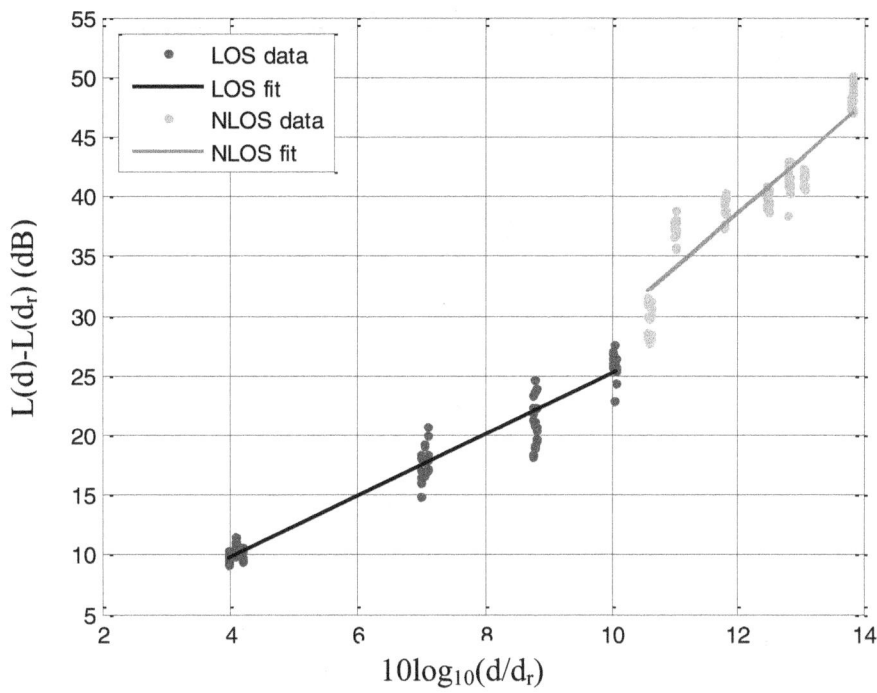

Figure 14. Excess propagation path loss (dB) vs. $10\log_{10}(\text{distance}/d_r)$, 700 MHz band.

Table 5. Path loss model parameters: n=path loss exponent, σ_X=standard deviation of Gaussian RV, L_c = corner loss.

		n	σ_X (dB)	L_c (dB)
700 MHz				
Tx1	**LOS**	2.57	1.46	4.23
	NLOS	4.57	2.13	
Tx2	**LOS**	2.34	2.94	8.27
	NLOS	5.76	2.23	
Tx3	**LOS**	4.37	2.63	11.69
	NLOS	3.42	3.44	
All Tx	**LOS**	2.27	3.06	
	NLOS	3.58	2.92	
4900 MHz				
Tx1	**LOS**	1.34	1.25	7.73
	NLOS	4.04	2.47	
Tx2	**LOS**	1.59	2.54	7.08
	NLOS	5.18	3.23	
Tx3	**LOS**	1.53	2.74	12.87
	NLOS	3.47	3.02	
All Tx	**LOS**	1.64	2.65	
	NLOS	3.35	3.16	

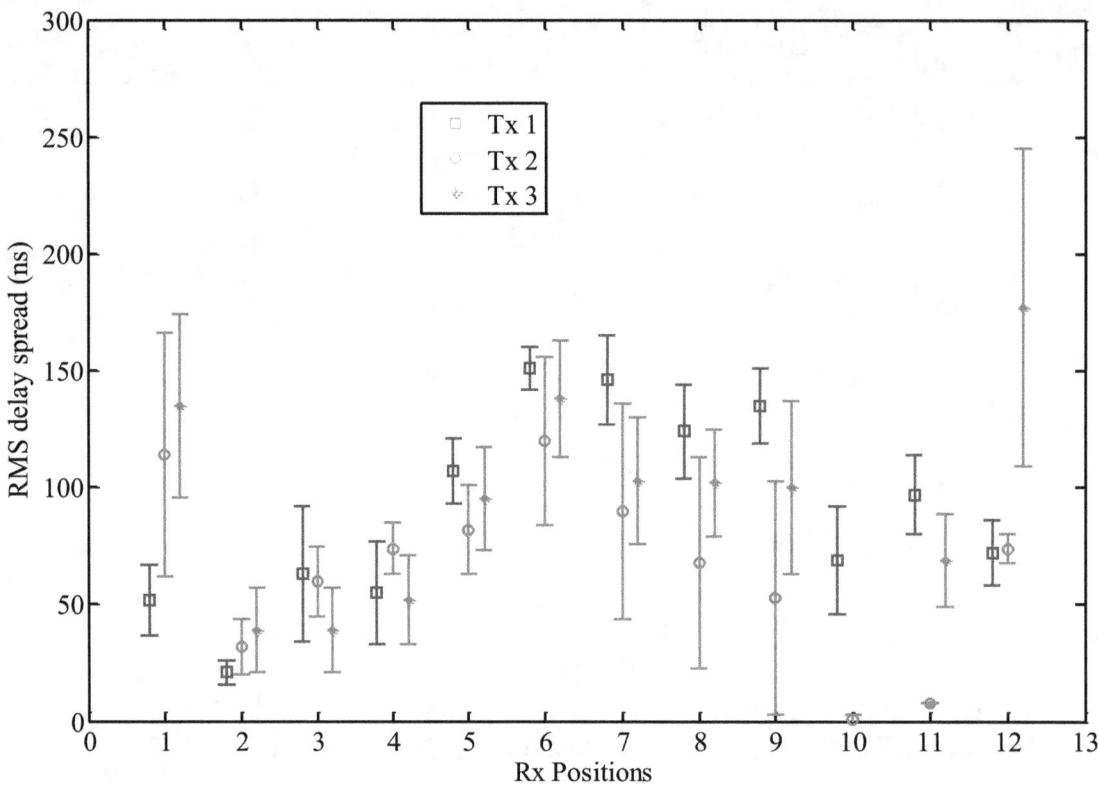

Figure 15. 700 MHz band, mean RMS delay spread ± σ for each receive antenna location for the first transmit location.

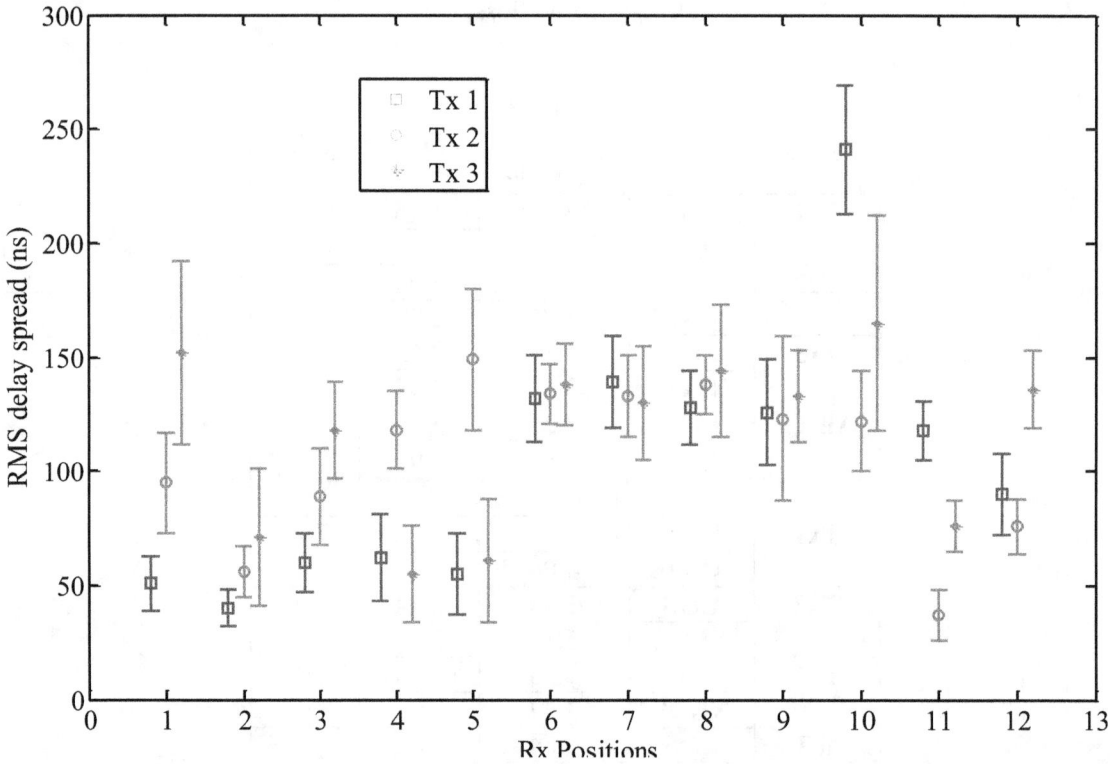

Figure 16. 4900 MHz band, mean RMS delay spread ± σ for each receive antenna location for second transmit location.

38

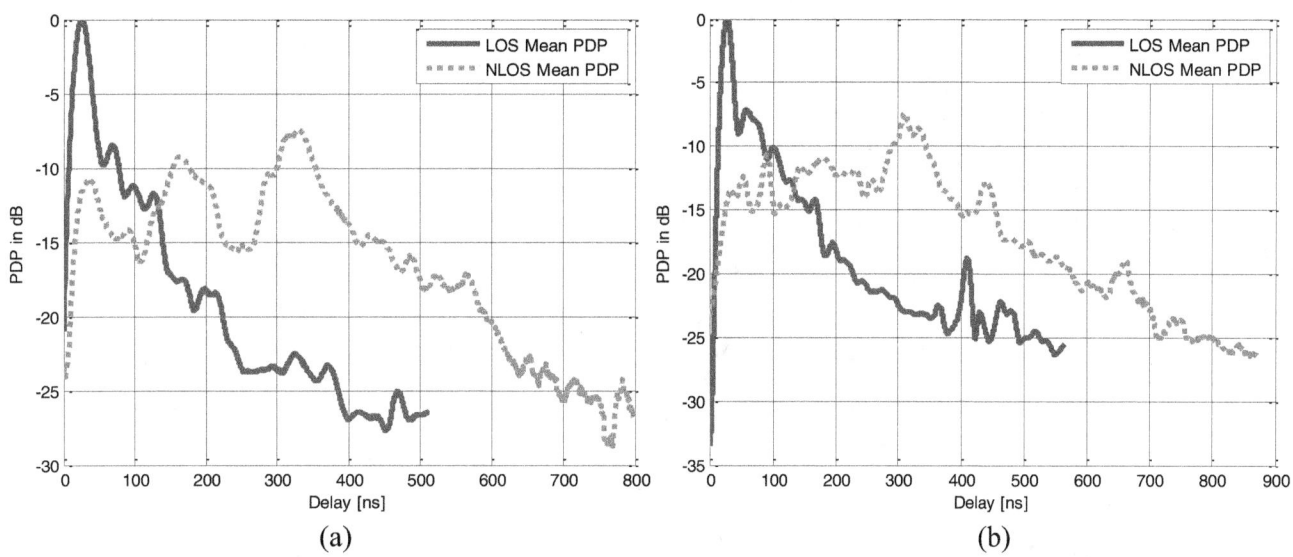

Figure 17. Average PDPs for all LOS and NLOS locations, (a) 700 MHz band; (b) 4900 MHz.

39

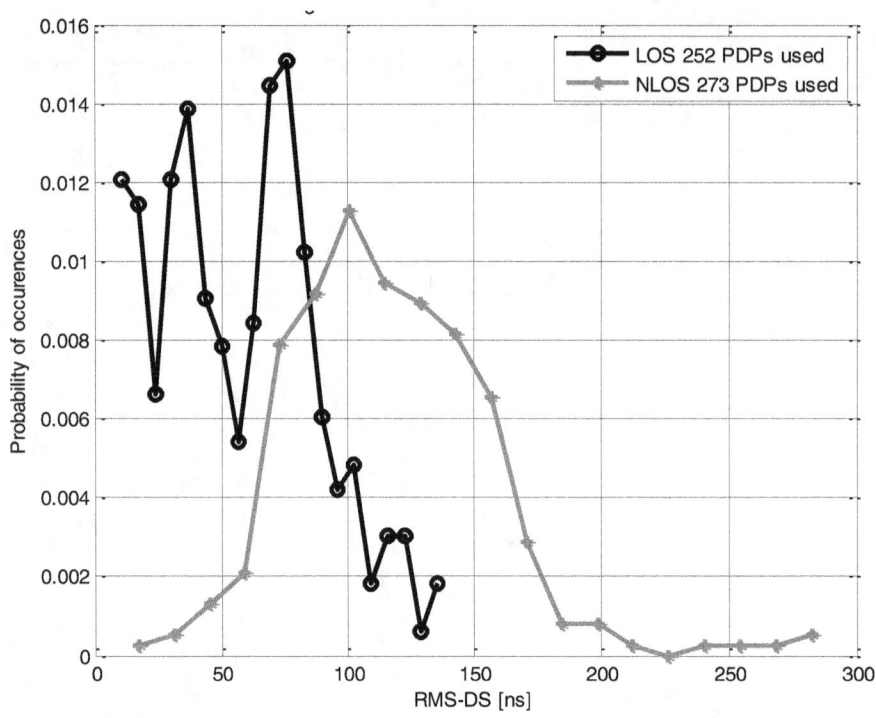

Figure 18. Histograms of RMS delay spread from power delay profiles in both LOS and NLOS locations, 700 MHz band.

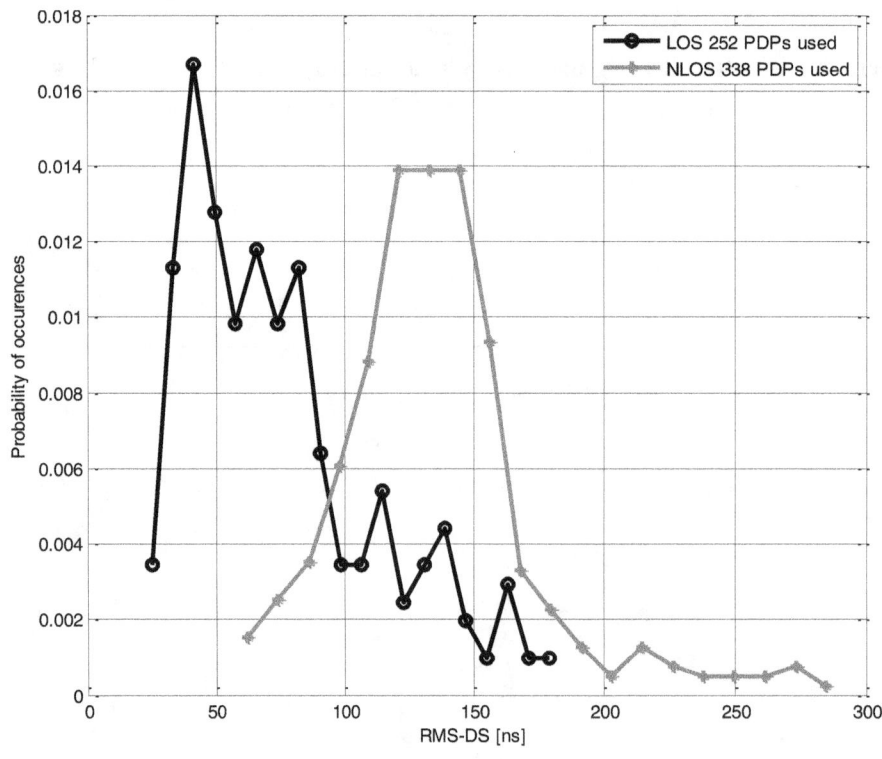

Figure 19. Histograms of RMS delay spread from power delay profiles in both LOS and NLOS locations, 4900 MHz band.

Table 6. Summary instantaneous RMS-DS statistics (ns).

Condition (band)	Min	Mean	Median	90th Percentile	Max	Standard Deviation
LOS (700)	7	57	56	**98**	**139**	32
LOS (4900)	21	75	66	**134**	**183**	37
NLOS (700)	11	116	114	**163**	**290**	40
NLOS (4900)	57	135	131	**172**	**291**	37

Table 7. 90% delay window W_{90} and 25 dB delay interval I_{25} statistics (ns).

Condition (band)	Min	Mean	Median	Max
LOS W_{90} (700)	24	152	133	427
LOS W_{90} (4900)	33	205	157	538
LOS I_{25} (700)	42	384	435	746
LOS I_{25} (4900)	112	466	469	970
NLOS W_{90} (700)	36	377	370	976
NLOS W_{90} (4900)	157	442	437	984
NLOS I_{25} (700)	51	645	671	1000
NLOS I_{25} (4900)	355	791	825	1000

Table 8. Statistics for number of multipath components L_p, with a 25 dB threshold from PDP peak, plus modified uniform probability mass function fit parameter p_0.

L_p Statistic	700 MHz		4900 MHz	
	LOS	NLOS	LOS	NLOS
Minimum	4	3	3	3
Median	11.50	13	17	20.50
Mean	10.49	11.47	16	17.91
90th Percentile	14	14	21	21
Maximum	14	14	21	21
Standard Deviation	3.34	3.10	4.70	4.74
p_0	0.3294	0.4428	0.2619	0.5

Table 9. PDP exponential fit parameters of equations (11), (12).

Parameter	700 MHz	4900 MHz
c_0	1.09	0.98
c_1	0.07	0.076
b_{10}	0.39	0.27
b_{11}	0.017	0.003
b_{20}	6.56	0.56
b_{21}	0.018	0.003
τ_2 (ns)	73	70
b_{30}	129	26.7
b_{31}	0.017	0.013
τ_3 (ns)	215	218

Table 10. multipath component delay distribution probability density function parameters for equations (13), (14).

Band	LOS Parameter v	NLOS Parameters (a,b)
700 MHz	318.2	(452.7, 1.57)
4900 MHz	340.6	(472.8, 1.6)

Appendix II: 700 MHz Experimental Data

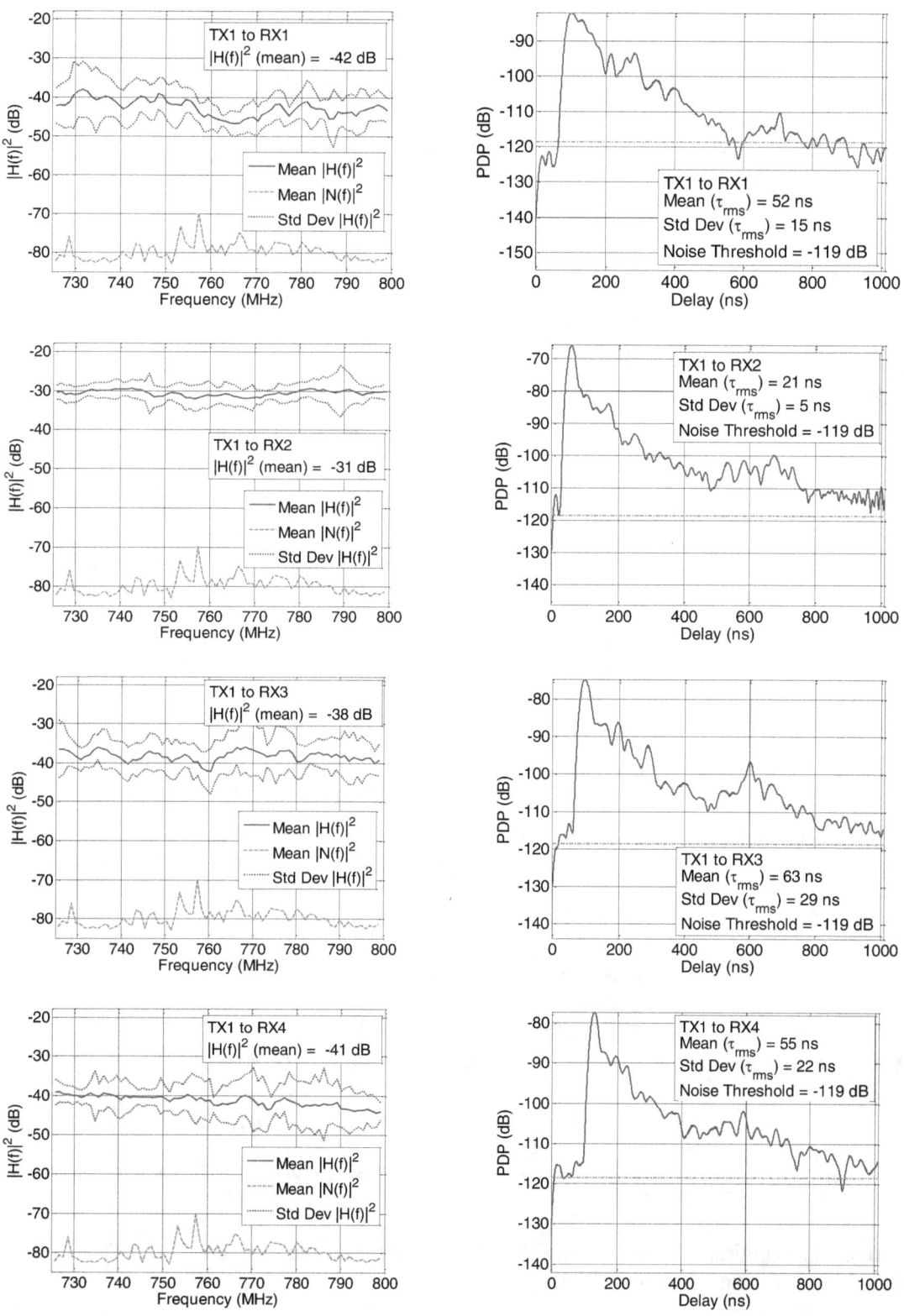

Figure 20. Magnitude squared of the channel performance and PDP for Tx1 to Rx1 through Rx4 pairings in the 700 MHz band.

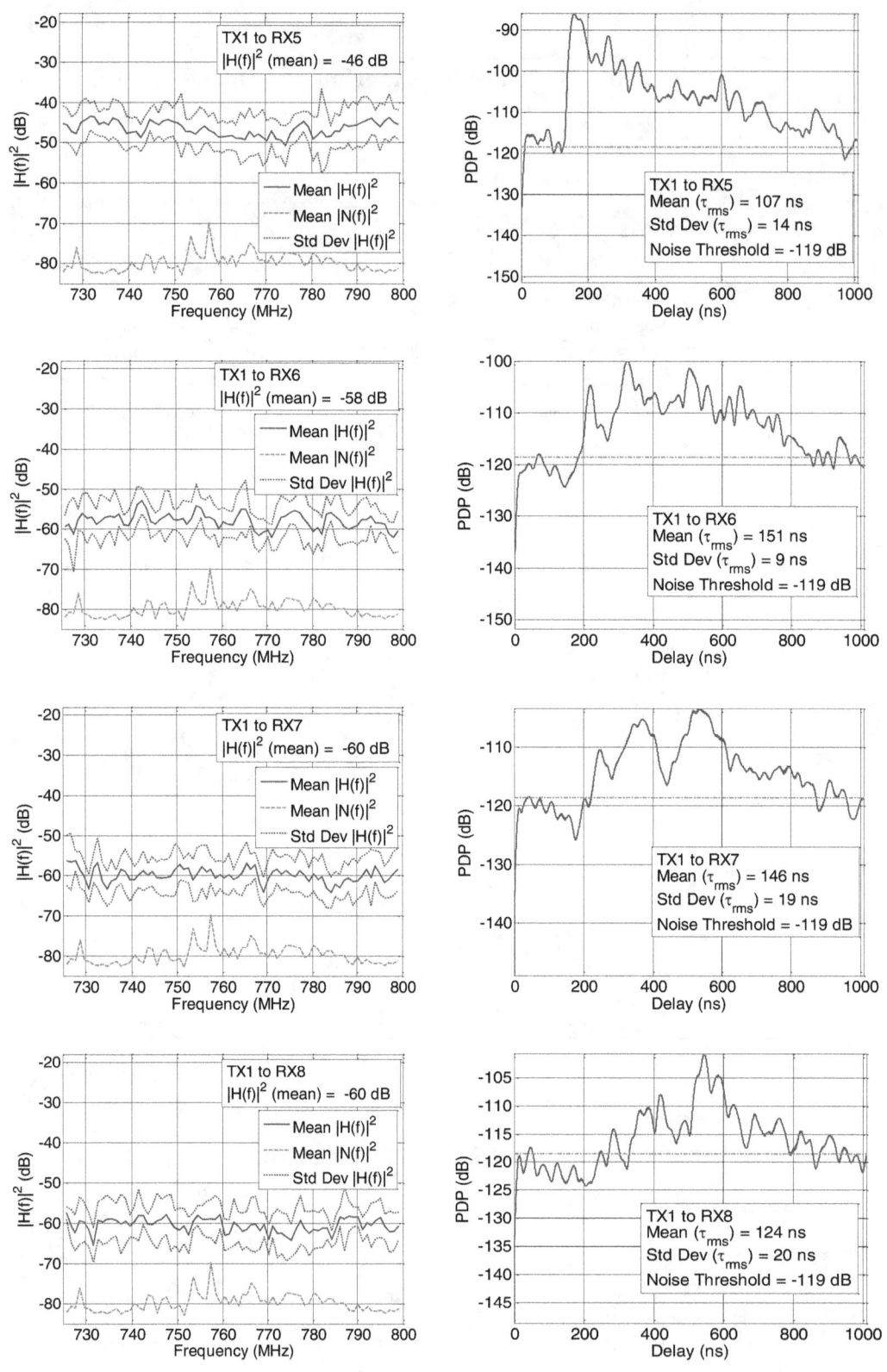

Figure 21. Magnitude squared of the channel performance and PDP for Tx1 to Rx5 through Rx8 pairings in the 700 MHz band.

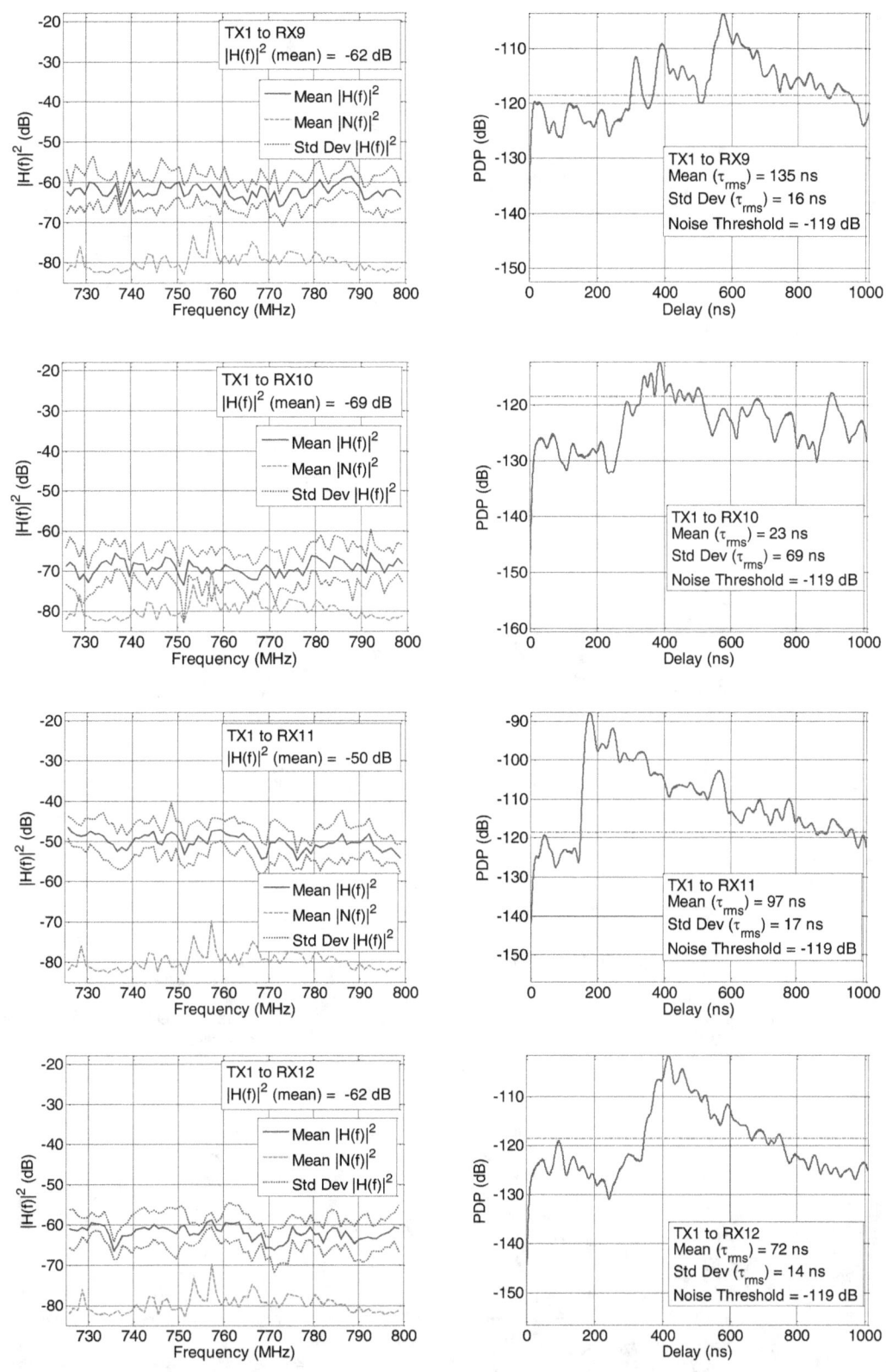

Figure 22. Magnitude squared of the channel performance and PDP for Tx1 to Rx9 through Rx12 pairings in the 700 MHz band.

45

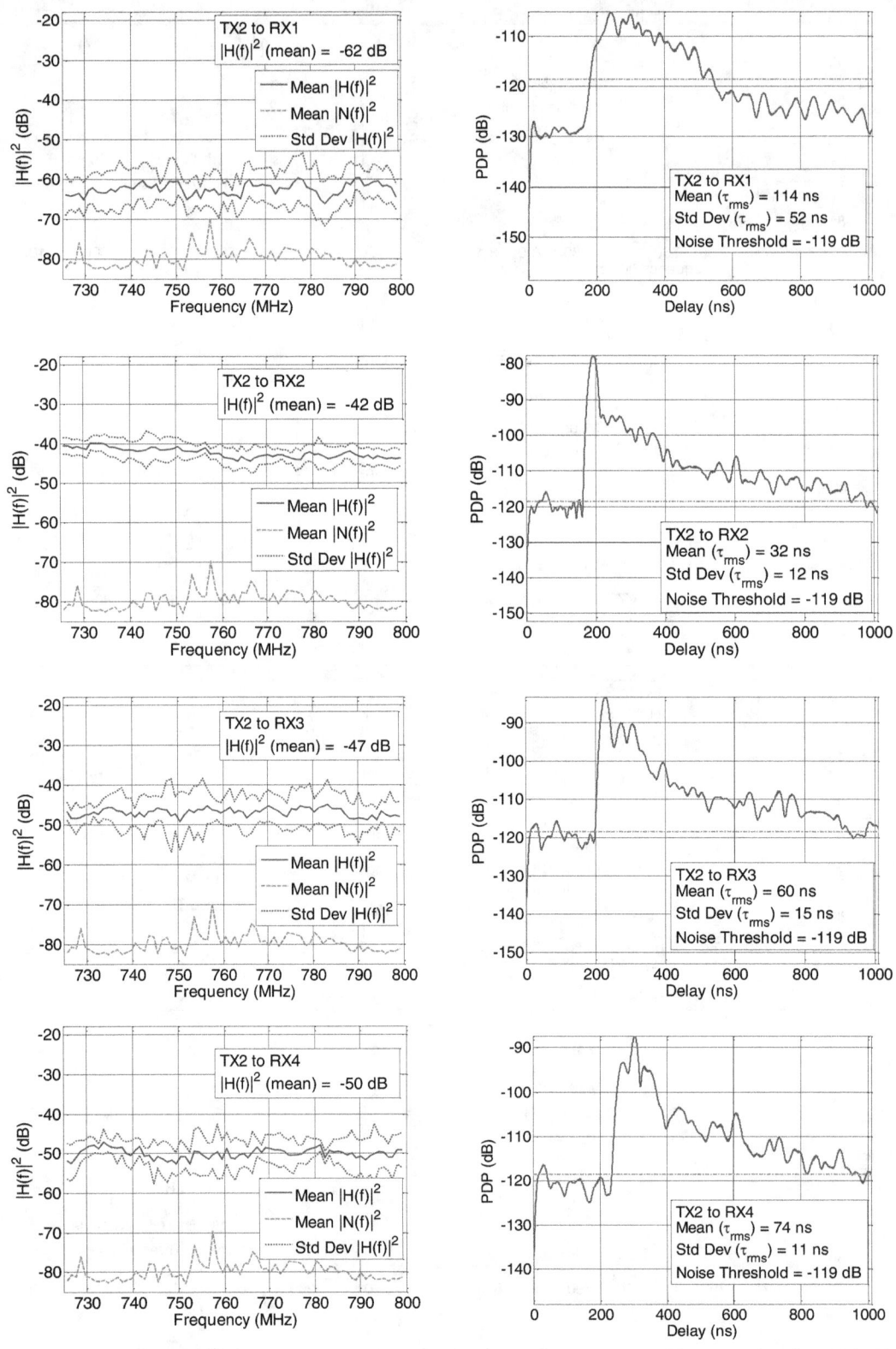

Figure 23. Magnitude squared of the channel performance and PDP for Tx2 to Rx1 through Rx4 pairings in the 700 MHz band.

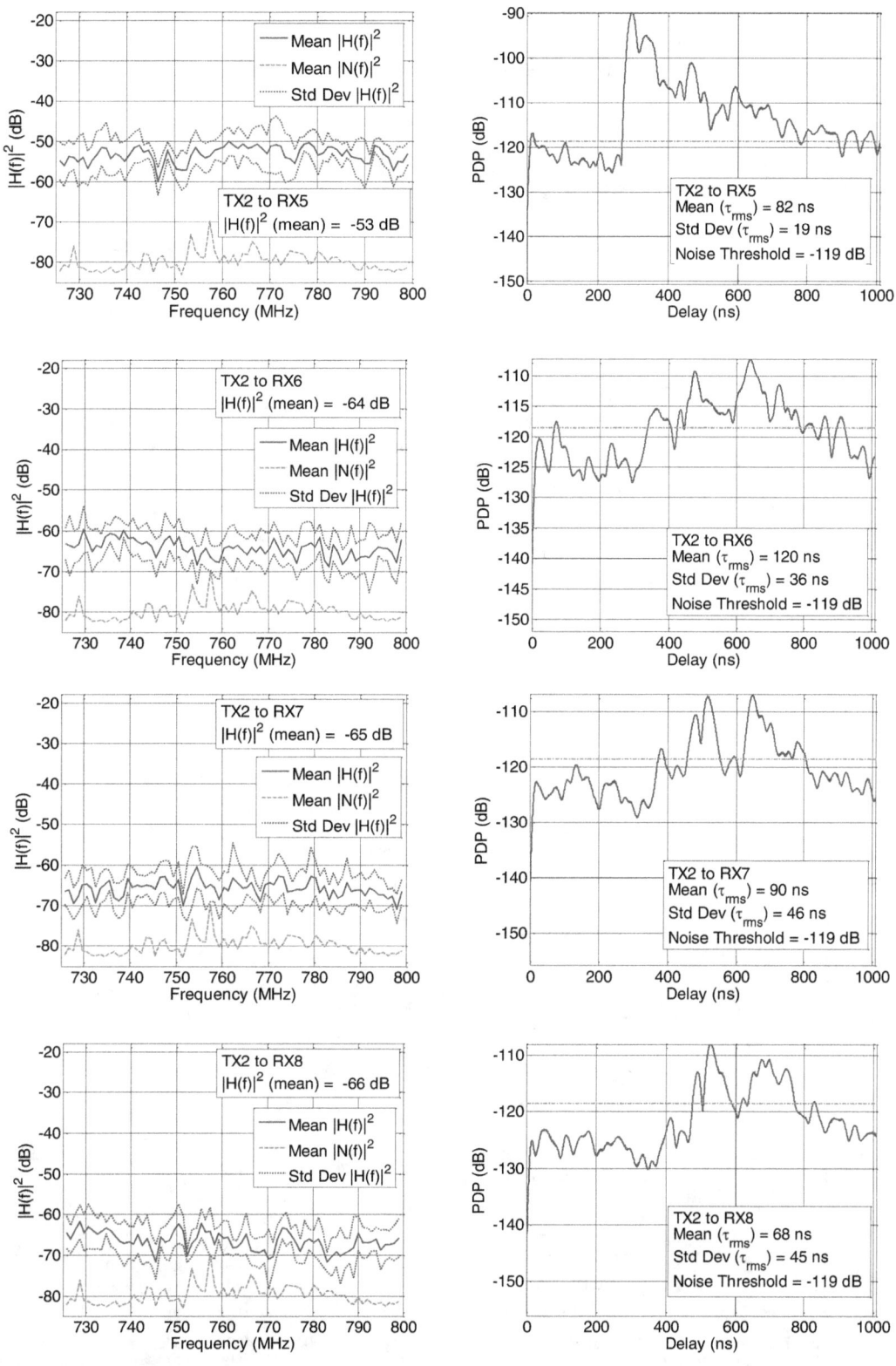

Figure 24. Magnitude squared of the channel performance and PDP for Tx2 to Rx5 through Rx8 pairings in the 700 MHz band.

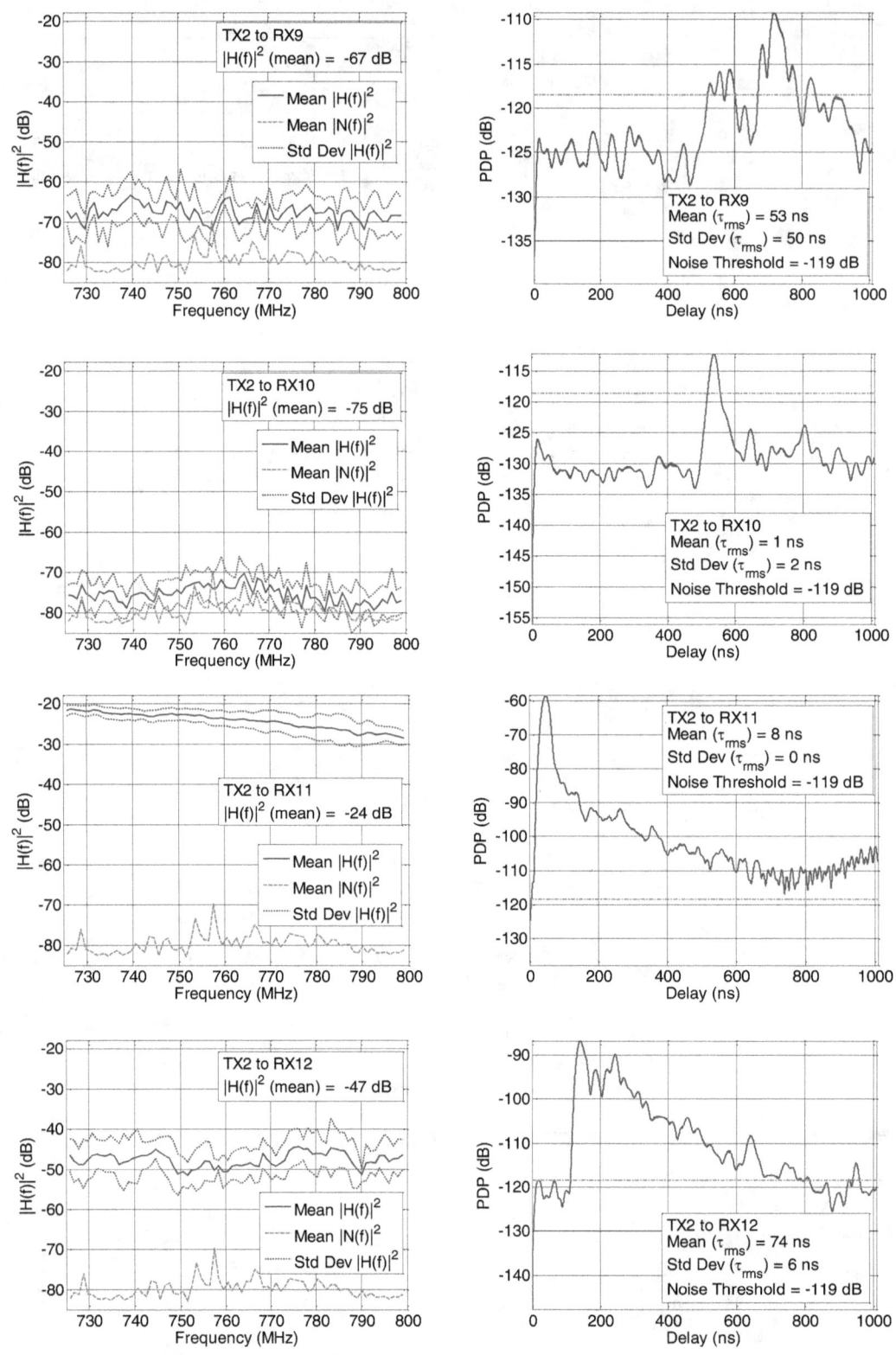

Figure 25. Magnitude squared of the channel performance and PDP for Tx2 to Rx9 through Rx12 pairings in the 700 MHz band.

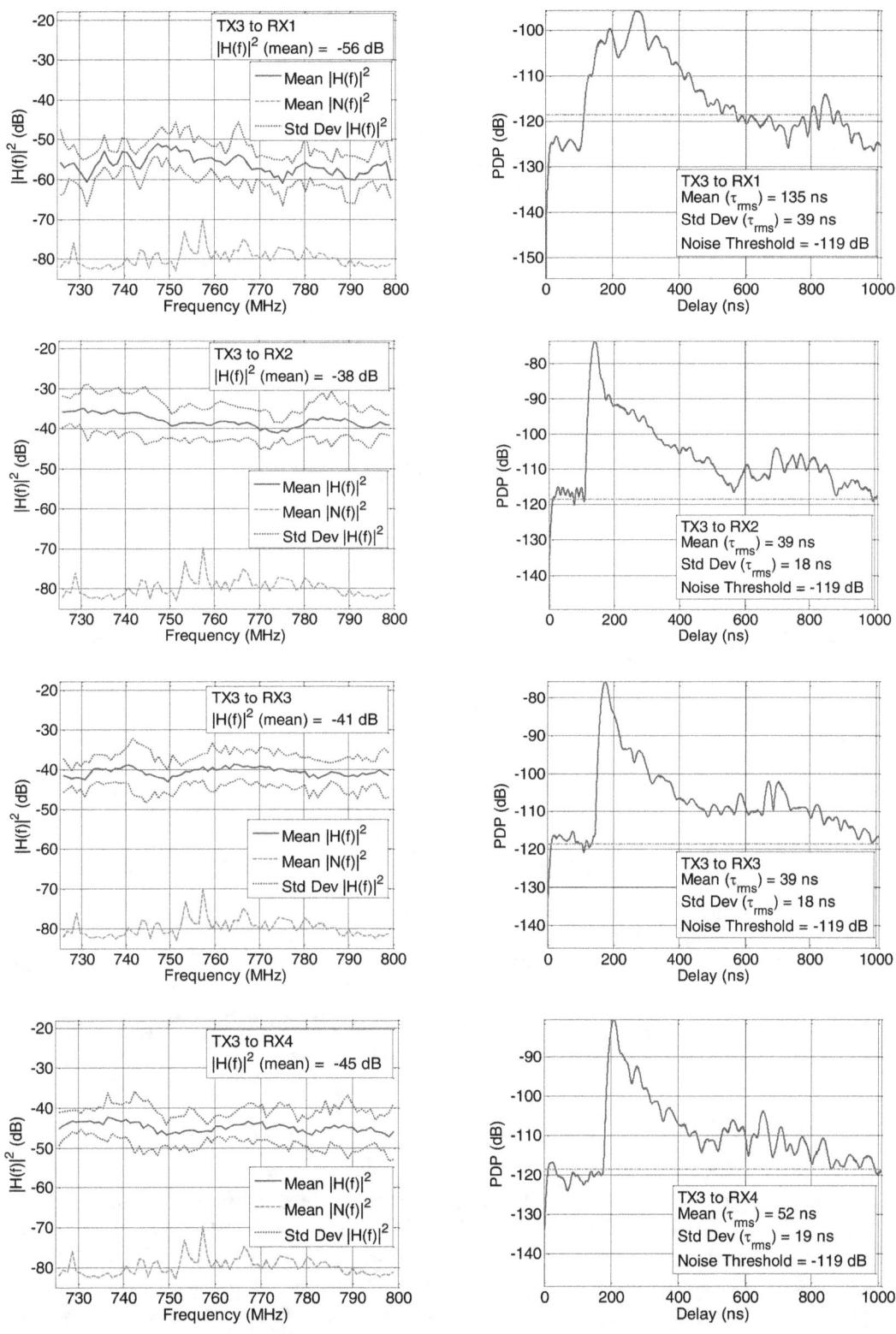

Figure 26. Magnitude squared of the channel performance and PDP for Tx3 to Rx1 through Rx4 pairings in the 700 MHz band.

49

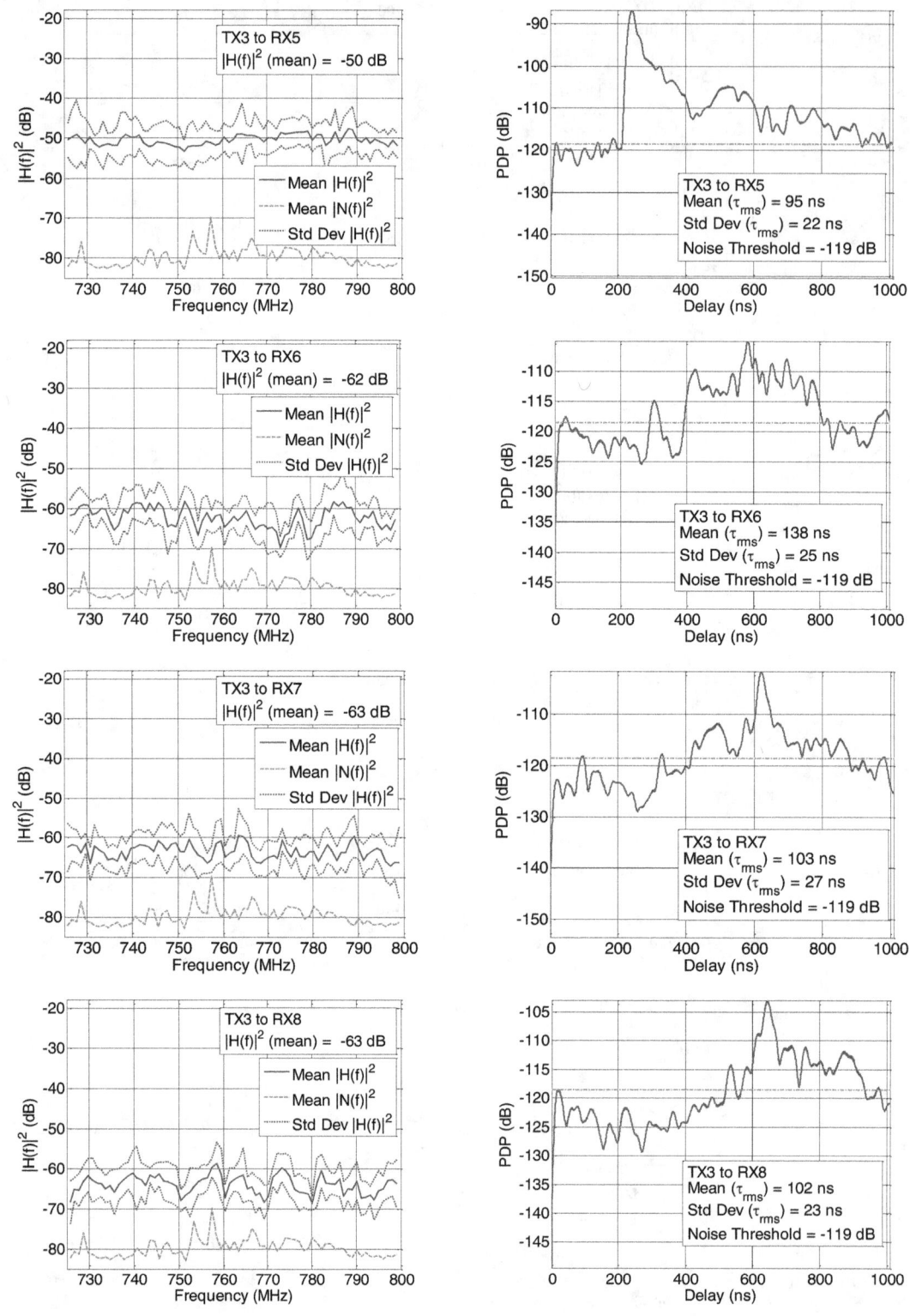

Figure 27. Magnitude squared of the channel performance and PDP for Tx3 to Rx5 through Rx8 pairings in the 700 MHz band.

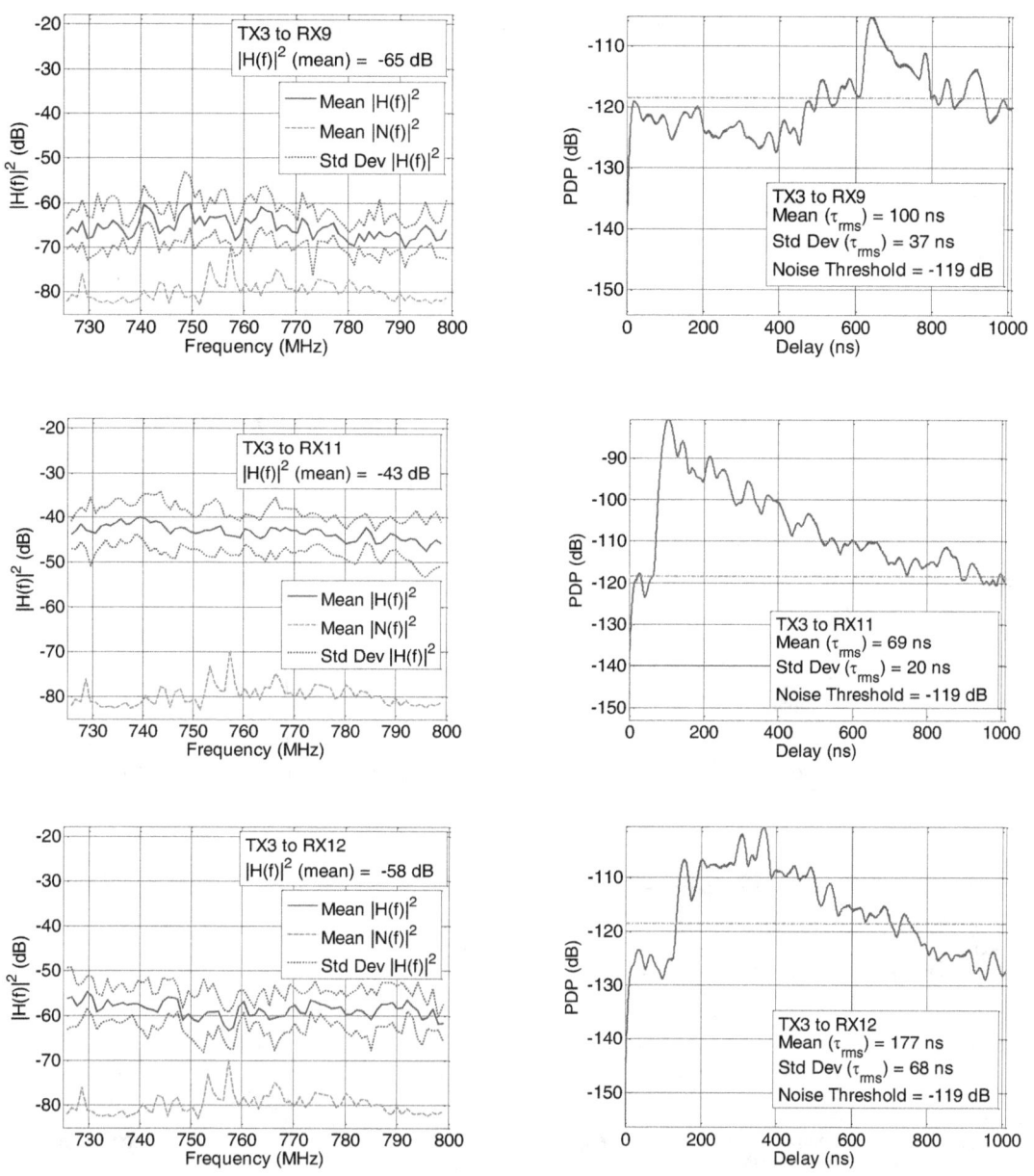

Figure 28. Magnitude squared of the channel performance and PDP for Tx3 to Rx9, Rx11, and Rx12 pairings in the 700 MHz band.

Appendix III: 4900 MHz Experimental Data

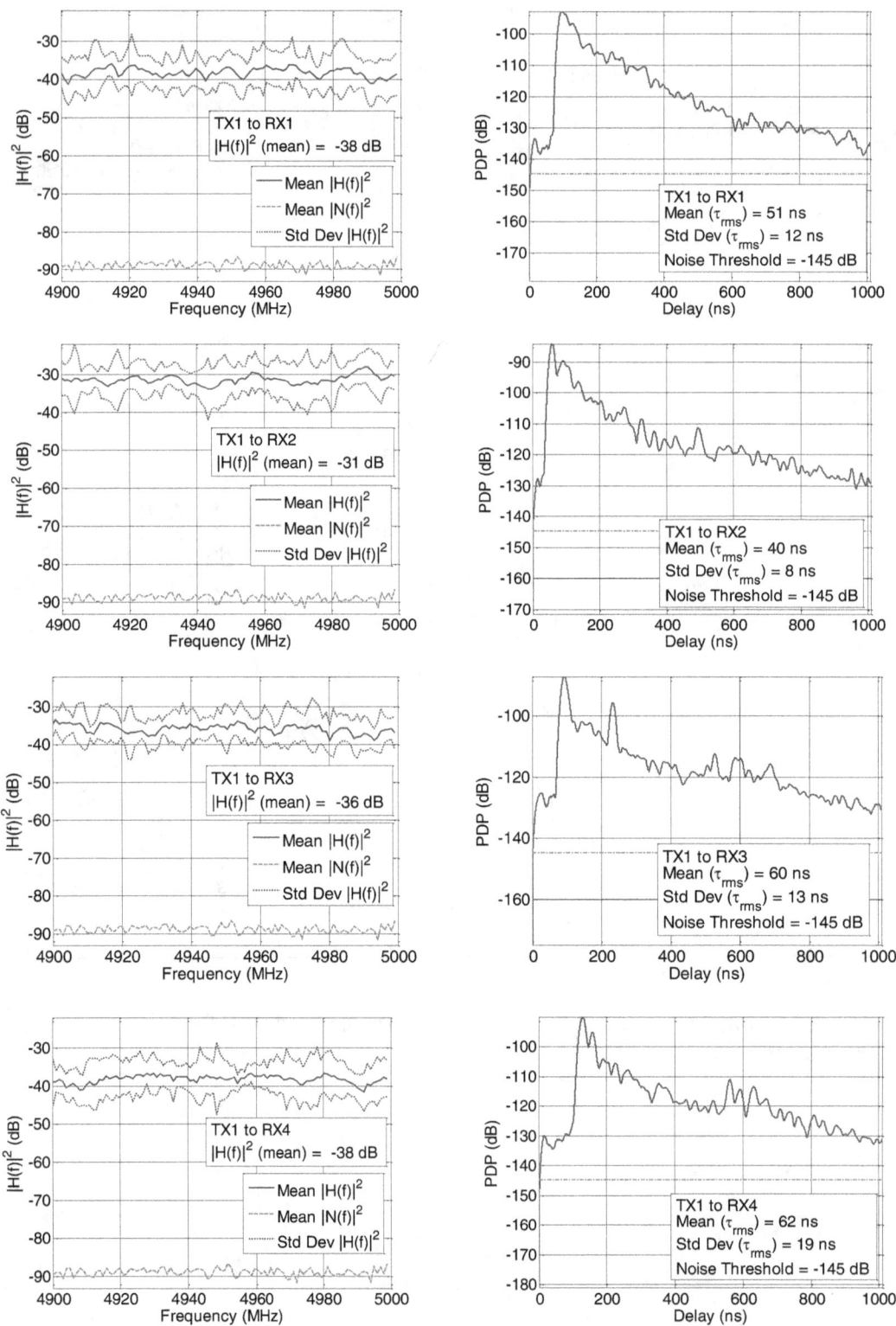

Figure 29. Magnitude squared of the channel performance and PDP for Tx1 to Rx1 through Rx4 pairings in the 4900 MHz band.

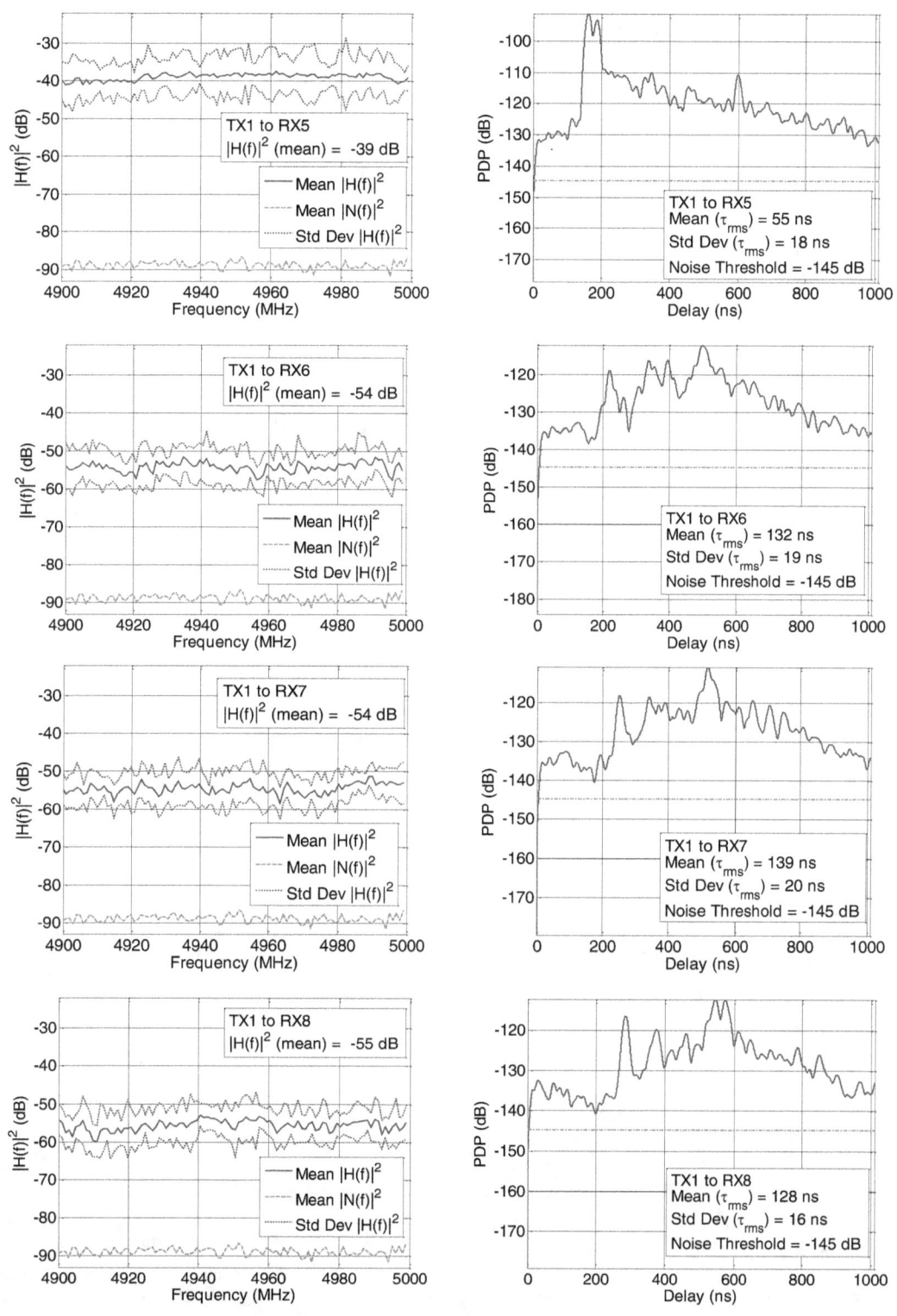

Figure 30. Magnitude squared of the channel performance and PDP for Tx1 to Rx5 through Rx8 pairings in the 4900 MHz band.

53

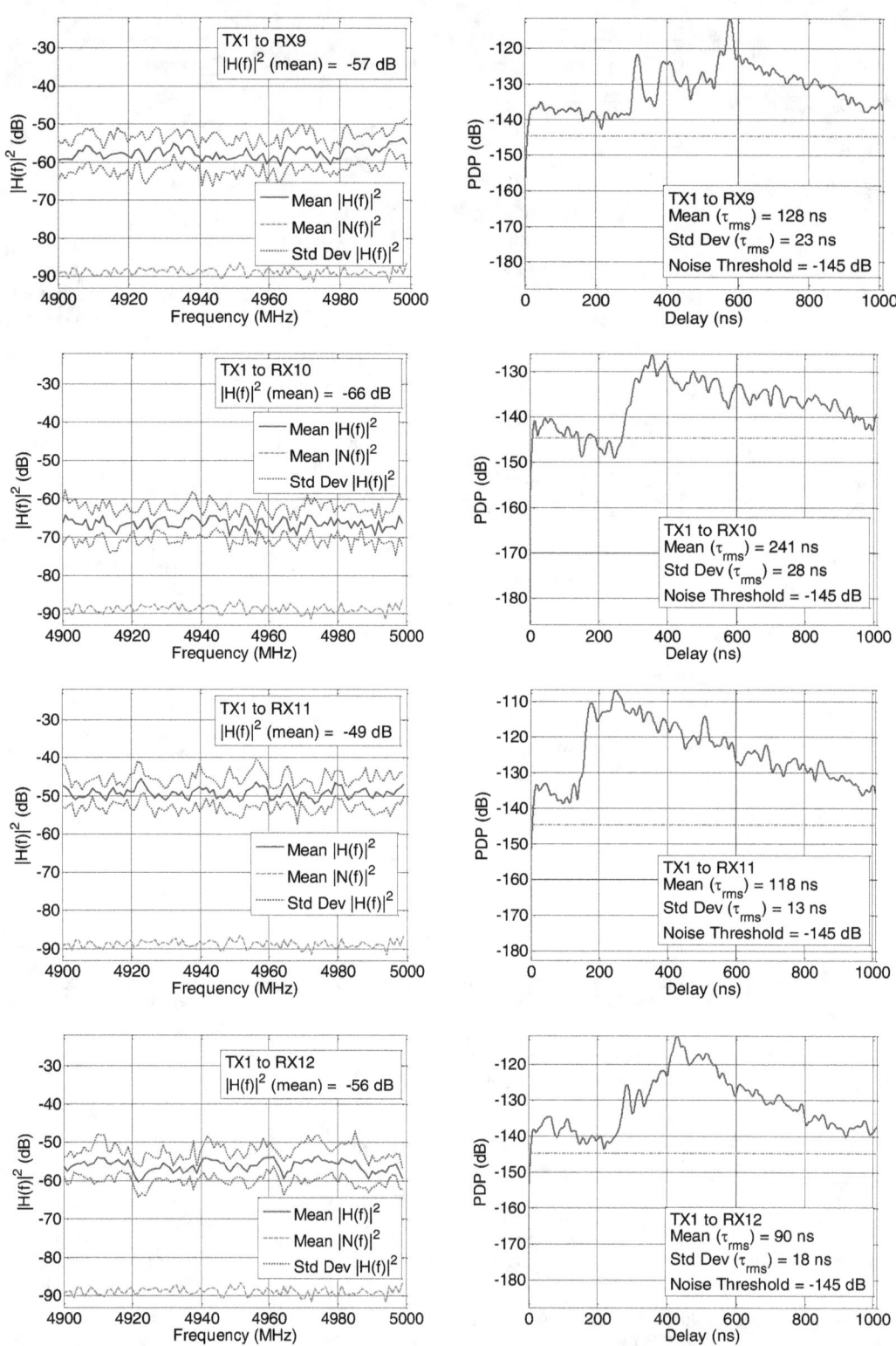

Figure 31. Magnitude squared of the channel performance and PDP for Tx1 to Rx9 through Rx12 pairings in the 4900 MHz band.

54

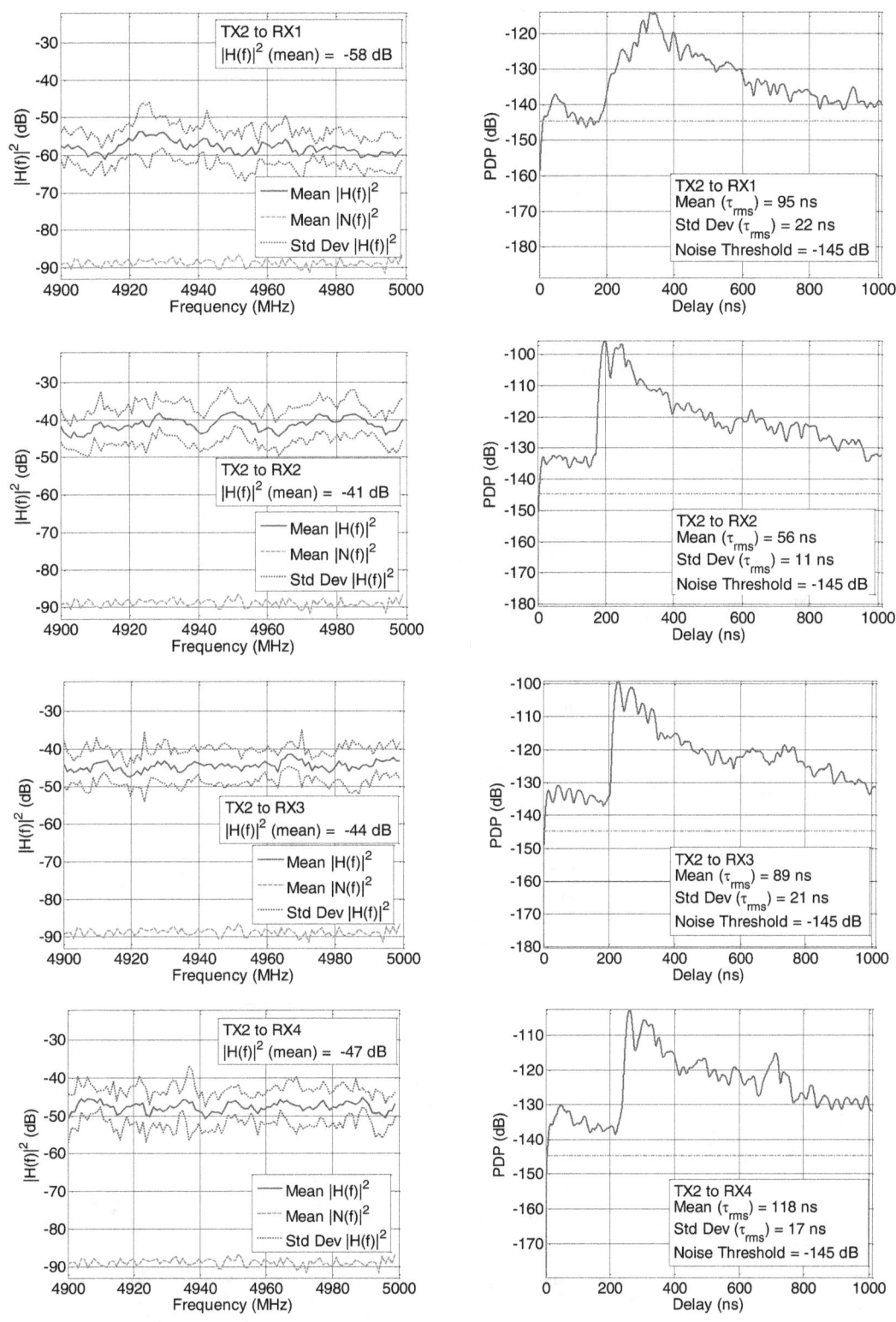

Figure 32. Magnitude squared of the channel performance and PDP for Tx2 to Rx1 through Rx4 pairings in the 4900 MHz band.

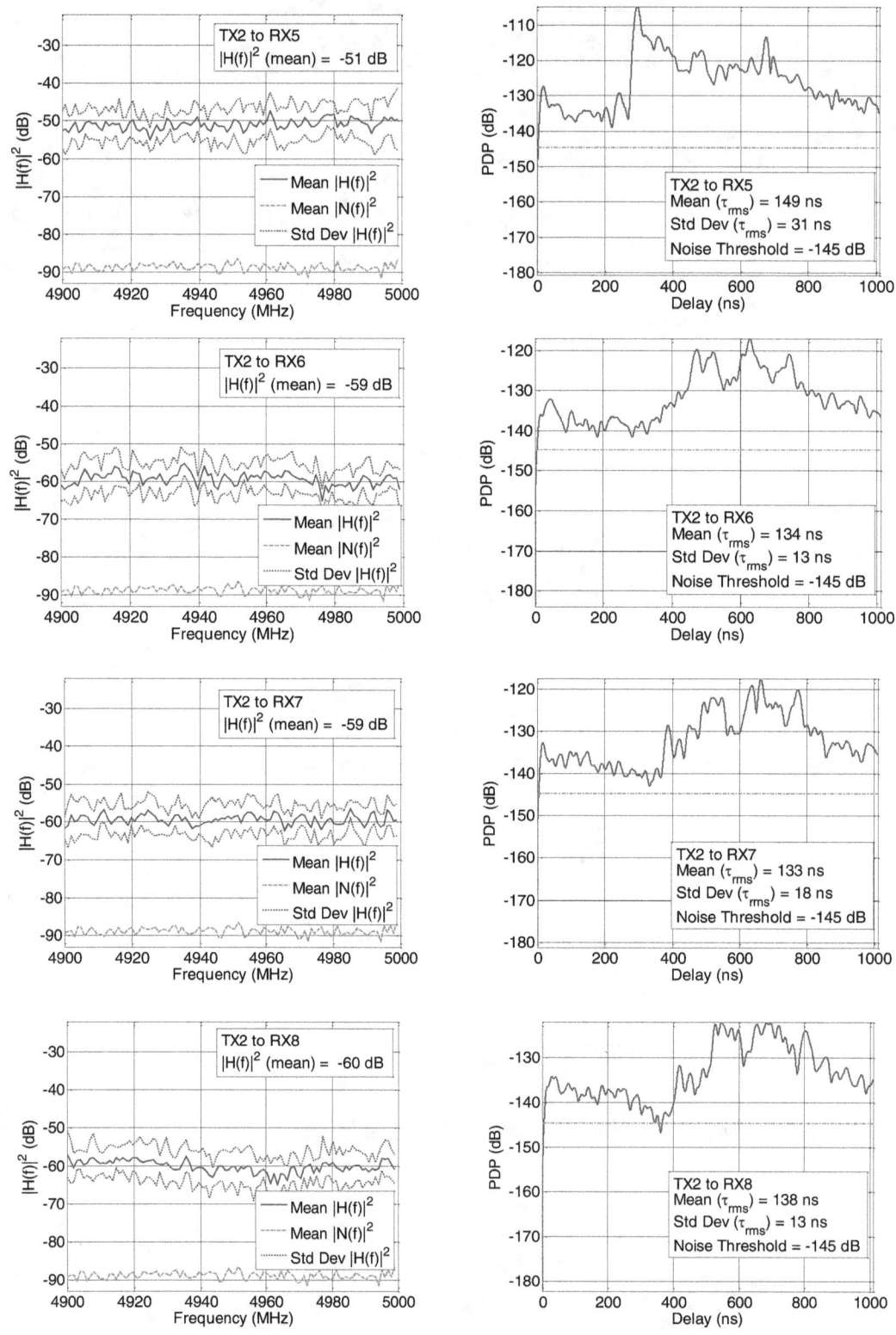

Figure 33. Magnitude squared of the channel performance and PDP for Tx2 to Rx5 through Rx8 pairings in the 4900 MHz band.

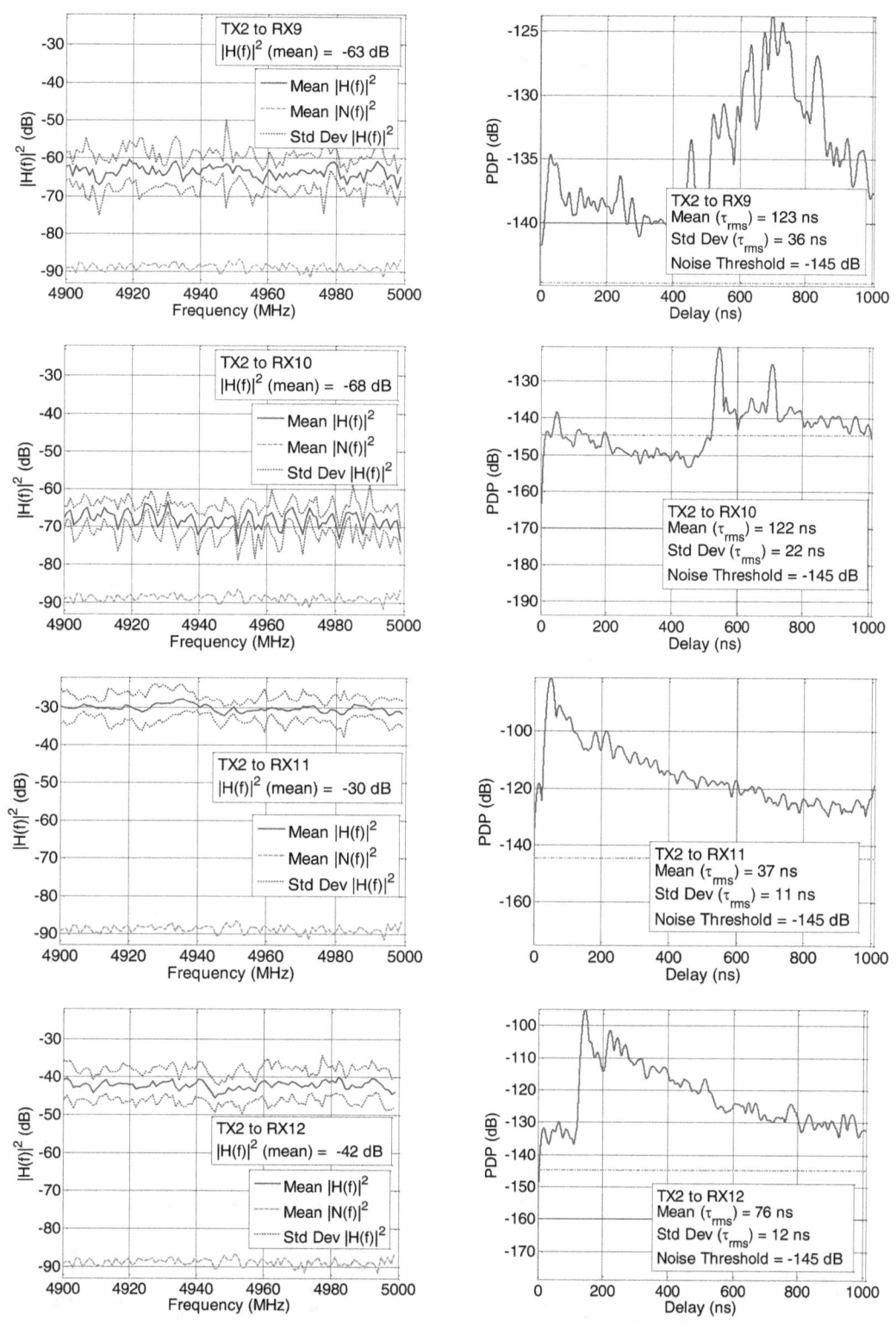

Figure 34. Magnitude squared of the channel performance and PDP for Tx2 to Rx9 through Rx12 pairings in the 4900 MHz band.

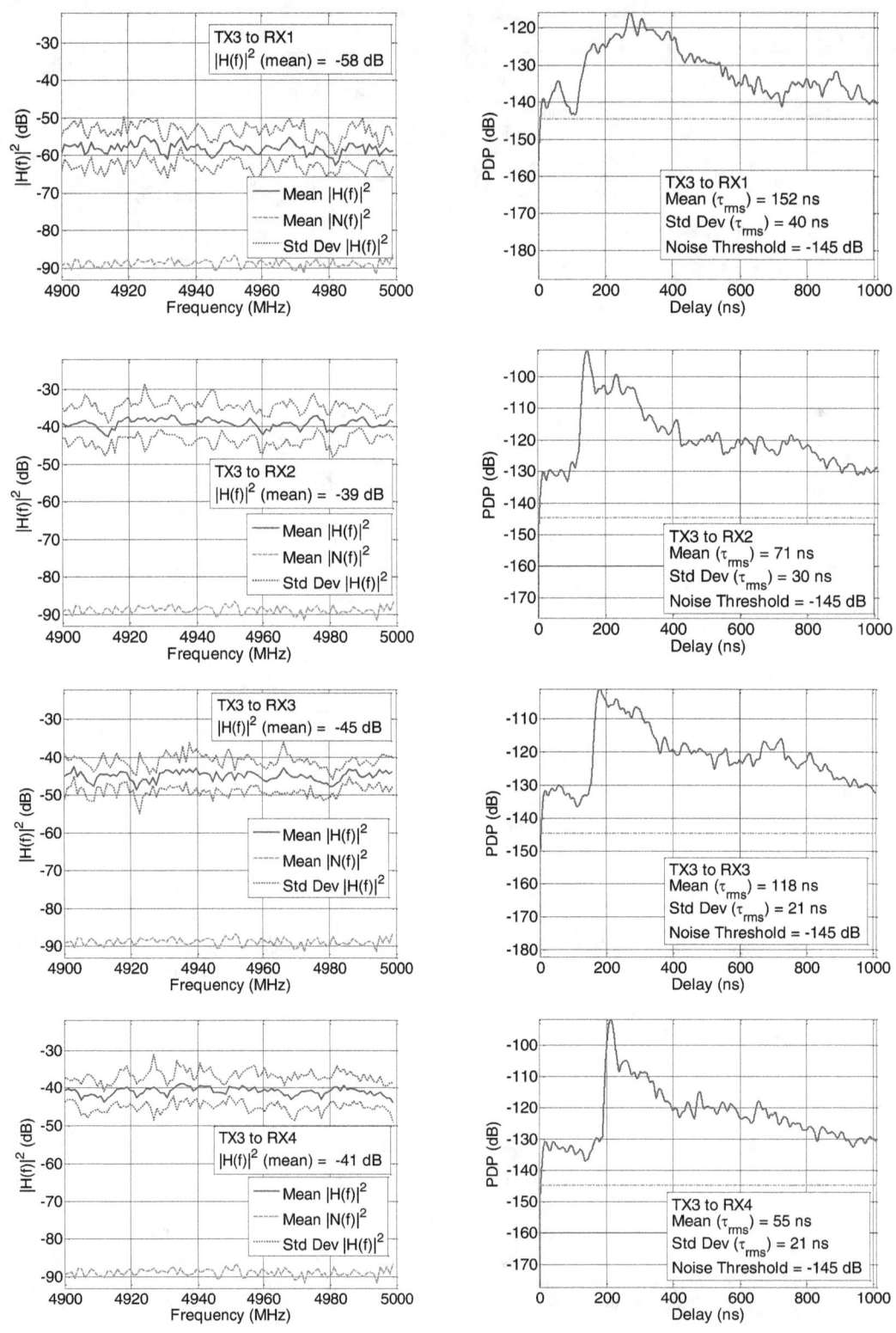

Figure 35. Magnitude squared of the channel performance and PDP for Tx3 to Rx1 through Rx4 pairings in the 4900 MHz band.

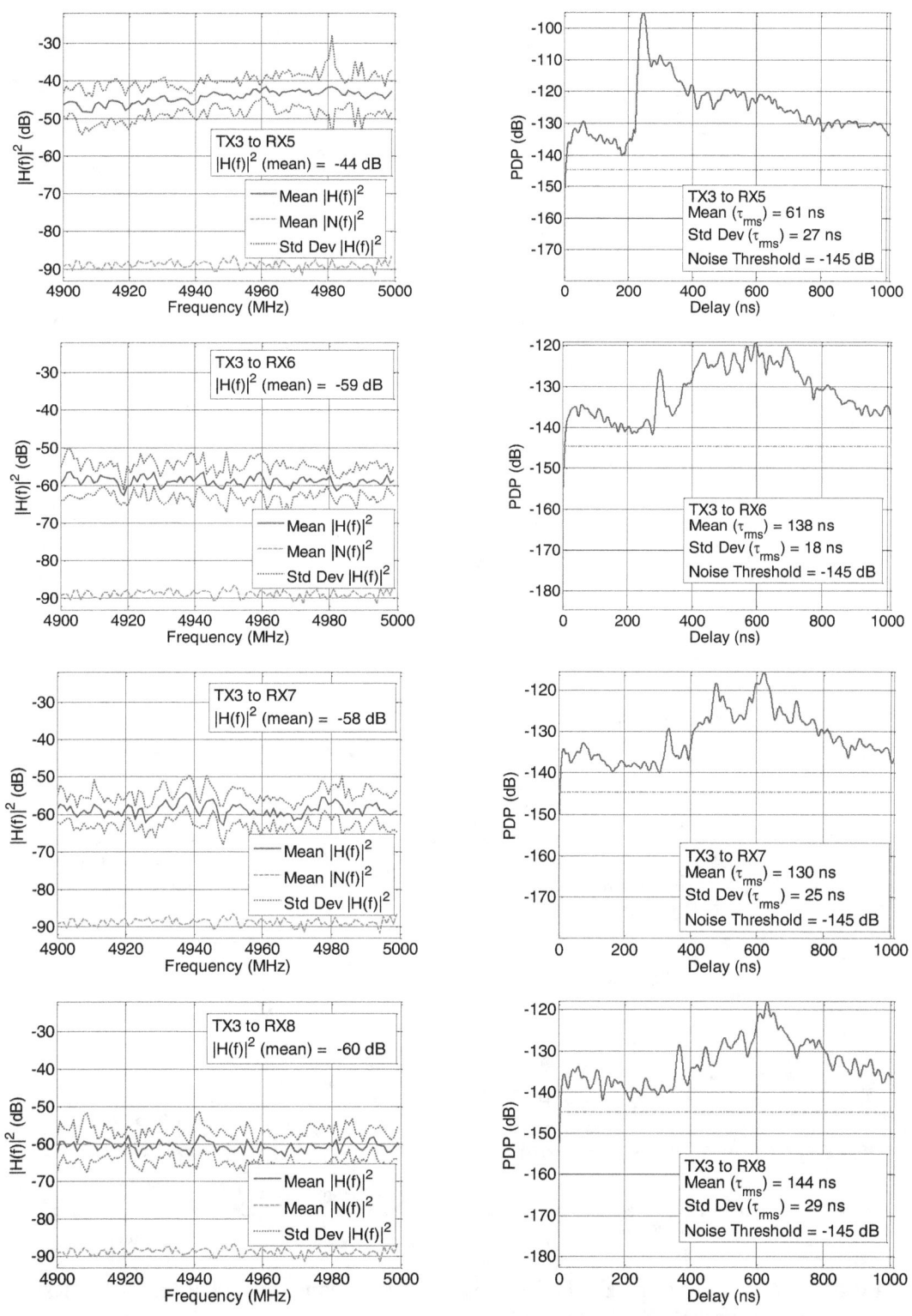

Figure 36. Magnitude squared of the channel performance and PDP for Tx3 to Rx5 through Rx8 pairings in the 4900 MHz band.

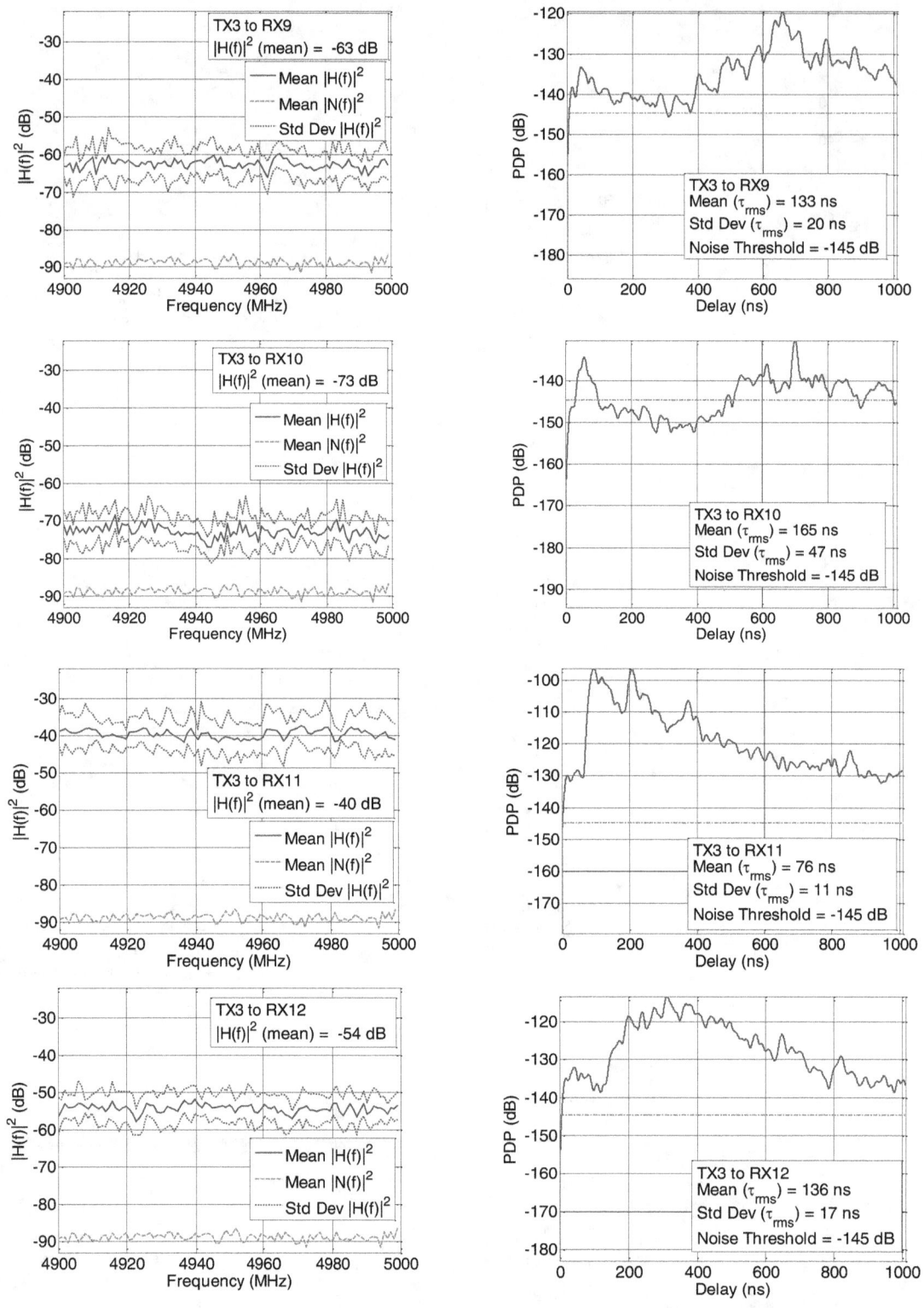

Figure 37. Magnitude squared of the channel performance and PDP for Tx3 to Rx9 through Rx12 pairings in the 4900 MHz band.

NIST *Technical Publications*

Periodical

Journal of Research of the National Institute of Standards and TechnologyCReports NIST research and development in metrology and related fields of physical science, engineering, applied mathematics, statistics, biotechnology, and information technology. Papers cover a broad range of subjects, with major emphasis on measurement methodology and the basic technology underlying standardization. Also included from time to time are survey articles on topics closely related to the Institute's technical and scientific programs. Issued six times a year.

Nonperiodicals

MonographsCMajor contributions to the technical literature on various subjects related to the Institute's scientific and technical activities.

HandbooksCRecommended codes of engineering and industrial practice (including safety codes) developed in cooperation with interested industries, professional organizations, and regulatory bodies.

Special PublicationsCInclude proceedings of conferences sponsored by NIST, NIST annual reports, and other special publications appropriate to this grouping such as wall charts, pocket cards, and bibliographies.

National Standard Reference Data SeriesCProvides quantitative data on the physical and chemical properties of materials, compiled from the world's literature and critically evaluated. Developed under a worldwide program coordinated by NIST under the authority of the National Standard Data Act (Public Law 90-396). NOTE: The Journal of Physical and Chemical Reference Data (JPCRD) is published bimonthly for NIST by the American Institute of Physics (AlP). Subscription orders and renewals are available from AIP, P.O. Box 503284, St. Louis, MO 63150-3284.

Building Science SeriesCDisseminates technical information developed at the Institute on building materials, components, systems, and whole structures. The series presents research results, test methods, and performance criteria related to the structural and environmental functions and the durability and safety characteristics of building elements and systems.

Technical NotesCStudies or reports which are complete in themselves but restrictive in their treatment of a subject. Analogous to monographs but not so comprehensive in scope or definitive in treatment of the subject area. Often serve as a vehicle for final reports of work performed at NIST under the sponsorship of other government agencies.

Voluntary Product StandardsCDeveloped under procedures published by the Department of Commerce in Part 10, Title 15, of the Code of Federal Regulations. The standards establish nationally recognized requirements for products, and provide all concerned interests with a basis for common understanding of the characteristics of the products. NIST administers this program in support of the efforts of private-sector standardizing organizations.

*Order the **following** NIST publicationsCFIPS and NISTIRsCfrom the National Technical Information Service, Springfield, VA 22161.*

Federal Information Processing Standards Publications (FIPS PUB)CPublications in this series collectively constitute the Federal Information Processing Standards Register. The Register serves as the official source of information in the Federal Government regarding standards issued by NIST pursuant to the Federal Property and Administrative Services Act of 1949 as amended, Public Law 89-306 (79 Stat. 1127), and as implemented by Executive Order 11717 (38 FR 12315, dated May 11,1973) and Part 6 of Title 15 CFR (Code of Federal Regulations).

NIST Interagency or Internal Reports (NISTIR)CThe series includes interim or final reports on work performed by NIST for outside sponsors (both government and nongovernment). In general, initial distribution is handled by the sponsor; public distribution is handled by sales through the National Technical Information Service, Springfield, VA 22161, in hard copy, electronic media, or microfiche form. NISTIRs may also report results of NIST projects of transitory or limited interest, including those that will be published subsequently in more comprehensive form.

U.S. Department of Commerce
National Bureau of Standards and Technology
325 Broadway
Boulder, CO 80305-3328

Official Business
Penalty for Private Use $300

)

www.ingramcontent.com/pod-product-compliance
Lightning Source LLC
Chambersburg PA
CBHW081852170526
45167CB00007B/2981